SMITHSONIAN CONTRIBUTIONS TO ZOOLOGY • NUMBER 559

The Intermuscular Bones and Ligaments of Teleostean Fishes

Colin Patterson and G. David Johnson

SMITHSONIAN INSTITUTION PRESS

Washington, D.C.

1995

ABSTRACT

Patterson, Colin, and G. David Johnson. The Intermuscular Bones and Ligaments of Teleostean Fishes. *Smithsonian Contributions to Zoology,* number 559, 85 pages, 16 figures, 2 plates, 8 tables, 1995.—Intermuscular bones are found only in teleostean fishes. They are segmental ossifications in the myosepta and generally are associated with ligaments. That association takes three forms: ontogenetic or structural continuity, when intermuscular bones ossify within ligament and/or are attached to the axial skeleton by ligament; serial homology, when a series of bones is continued rostrally or caudally by a series of ligaments; and homology, when a series of bones in one teleost is homologous with a series of ligaments in another. We recognize three series of intermusculars, epineurals, epicentrals, and epipleurals. Epineurals and epicentrals develop in a rostrocaudal gradient, whereas epipleurals develop rostrally and caudally from the region of the first caudal vertebra. We create a notation for recording the distribution and form of intermuscular bones and ligaments, and we map them in about 125 genera of teleosts from over 100 families, covering all major groups. The primitive state of Recent teleostean intermusculars is exemplified by *Hiodon,* in which all ossified epineurals are fused with the neural arches, and all epicentrals and epipleurals are ligaments. Some or all epineural bones are free (unfused) from the neural arches in other teleosts, and in many lower (nonacanthomorph) elopocephalans they develop an anteroventral branch so that they are forked proximally. Epineurals are primitively dorsolaterally directed, but the first one to three are deflected ventrally in a few nonacanthomorphs (argentinoids, some aulopiforms, *Neoscopelus*) and in lampridiform acanthomorphs. In *Polymixia,* the first epineural is displaced ventrally into the horizontal septum, and in all other acanthomorphs several or all epineurals are so displaced; the bones generally called epipleurals in acanthomorphs are epineurals. Epicentrals lie in the horizontal septum and are primitively ligamentous. There are ossified epicentrals in *Notopterus, Megalops,* clupeiforms, gonorynchiforms, gymnotoids, *Thymallus,* and the aulopiforms *Parasudis, Alepisaurus,* and *Omosudis.* Epicentral ligaments sometimes include a cartilage rod distally (salmonoids, osmeroids, *Maurolicus, Polymixia*), and in many clupeoids the distal tip of each anterior epicentral bone is associated with a separate superficial chevron of cartilage. Anterior epicentrals are lacking in some aulopiforms and among acanthomorphs in beryciforms, some zeiforms, and primitive percomorphs, so that the series of ligaments begins on the posterior abdominal vertebrae. All epicentrals are absent in some aulopiforms, in all examined paracanthopterygians and stephanoberyciforms, and in many percomorphs. A series of segmental, anterolaterally directed ligaments, "POTs," attaches to epicentrals in the horizontal septum of many teleosts. In percomorphs and zeiforms the anterior POTs acquire a new association, attaching to epineural bones secondarily positioned in the horizontal septum. Epipleurals lie below the horizontal septum and are posteroventrally directed. There are ossified epipleurals in *Heterotis,* elopomorphs, clupeomorphs, esocoids, ostariophysans, argentinoids, stomiiforms, aulopiforms, myctophiforms, and *Polymixia.* Like the epineurals, in many lower (nonacanthomorph) elopocephalan teleosts the epipleural bones develop an anterodorsal branch so that they are forked proximally. Epipleurals are unossified in salmonoids and osmeroids (except *Spirinchus* and some galaxiines), and this, together with cartilaginous epicentrals in those groups, indicates that they are sister groups. Aulopiforms are uniquely characterized by attached epipleurals that extend forward to the first or second vertebra. In many aulopiforms, the anterior epipleurals are displaced dorsally into the horizontal septum, the reverse of the acanthomorph situation (epineurals displaced ventrally), and in most aulopiforms the primitive bidirectional pattern of epipleural development is replaced by a rostrocaudal gradient. The epipleural series is lost in all acanthomorphs except *Polymixia* and holocentrids. The distribution and structure of ribs and Baudelot's ligament also are mapped. The potential systematic value of the intermusculars is illustrated by a parsimony analysis of Aulopiformes, and intermuscular characters of many other groups are enumerated.

OFFICIAL PUBLICATION DATE is handstamped in a limited number of initial copies and is recorded in the Institution's annual report, *Smithsonian Year.* SERIES COVER DESIGN: The coral *Montastrea cavernosa* (Linnaeus).

Library of Congress Cataloging-in-Publication Data
Patterson, Colin
The intermuscular bones and ligaments of teleostean fishes / Colin Patterson and G. David Johnson.
p. cm. — (Smithsonian contributions to zoology ; no. 559)
Includes bibliographical references (p. 76).
1. Osteichthyes—Anatomy. 2. Bones. 3. Ligaments. 4. Muscles. I. Johnson, G. David. II. Title. III. Title: Teleostean fishes. IV. Series.
QL1.S54 no. 559 [QL639] 591 s—dc20 [597'.50447] 94-10700

∞ The paper used in this publication meets the minimum requirements of the American National Standard for Permanence of Paper for Printed Library Materials Z39.48—1984.

Contents

The Intermuscular Bones and Ligaments of Teleostean Fishes

Colin Patterson and G. David Johnson

"In studying the skeleton, we generally give too little attention to the ligaments. The oversight is regrettable. When two organic systems have such close mutual relations as the bones and the ligaments, and especially when the elements of one system may happen to take on the characters of the other, to study one without taking account of the other will often create the most serious difficulties. In morphology there is no such thing as insignificant detail; each observation, no matter how trifling, may carry the germ of an explanation for others of much greater consequence."

Émile Baudelot (1868:84, our translation from the French)

Introduction

The work reported here began with an attempt by Patterson to understand the intermuscular bones and ligaments in *Polymixia,* the genus currently regarded as the sister taxon of all other Acanthomorpha (Rosen, 1985; Stiassny, 1986). Alone among Recent acanthomorphs, *Polymixia* has been described as having two sets of intermuscular bones, epineurals and epipleurals (Starks, 1904a; Patterson, 1964; Zehren, 1979). Other Recent acanthomorphs have either a single series of bones, generally called epipleurals, or they lack intermuscular bones (bothid and samarid pleuronectiforms and aulostomid gasterosteiforms are exceptional in having extra autapomorphous series of bones; Amaoka, 1969; Hensley and Ahlstrom, 1984; Jungersen, 1910:270; and below). The work expanded into the present collaboration after Johnson saw a draft manuscript by Patterson and thereupon pointed out that intermuscular ligaments are much more widely distributed than Patterson had realized. We began a wider survey of teleostean intermuscular bones and ligaments and soon found that interpreting and recording details of the form and distribution of those structures in cleared-and-stained specimens can be extremely difficult and is best undertaken by two people, alternating as observer and as critic and recorder. The plan of the paper is first to consider historical and current terminology, next to give a detailed description of conditions in *Polymixia,* and then to summarize our observations and interpretations in separate sections on lower (nonacanthomorph) and acanthomorph teleosts. We recognize that the results presented here do no more than scratch the surface of a very complex subject; we believe that the intermuscular bones and ligaments hold promise as a source of characters of systematic value, and we hope that others will pursue and realize that promise.

Colin Patterson, The Natural History Museum, London, SW7 5BD, England. G. David Johnson, National Museum of Natural History, Smithsonian Institution, Washington, DC 20560.

Reviewers: Victor G. Springer, Smithsonian Institution; Gareth Nelson and Melanie J. Stiassny, The American Museum of Natural History; George Lauder, The University of California, Irvine.

Terminology and Material

Intermuscular bones, which occur only in teleosts amongst Recent vertebrates, are segmental, serially homologous ossifications in the myosepta. Their position in the myosepta distinguishes them from (pleural) ribs, which (with a few exceptions) are found not in myosepta but in the peritoneal membrane. In this paper we concentrate on the three principal series of intermusculars—epineurals, epicentrals, and epipleurals—that attach to the axial skeleton and which we show to be homologous throughout teleosts. In a few teleosts there are two or more additional series of intermusculars, the myorhabdoi (Chapman, 1944), which are commonest in the dorsal and ventral forward flexures of the myoseptum; however, myorhabdoi do not attach to the axial skeleton and are autapomorphous for those taxa in which they occur.

The terms "epineural," "epicentral," and "epipleural" were introduced by Richard Owen. "Epipleural" dates from 1846

(Owen, 1846:66), for bones "attached to, or near to, the heads of the ribs, [which] extend upward, outward, and backward, between the dorsal and lateral masses of muscles." "Epineural" and "epicentral" were introduced in 1866 (Owen, 1866:43), and this division of the intermusculars into three series generally has been followed since. Owen's "type locality" for the three series was the herring, *Clupea* (Owen, 1866, fig. 37), which has epineurals above the horizontal septum, epipleurals below it, and epicentrals in it. Owen did not mention the horizontal septum as a criterion, but he named the three series "according to the vertebral element they may adhere to" (Owen, 1866:43), i.e., neural arch or spine for the epineural, centrum for the epicentral, and rib for the epipleural. But he acknowledged that this criterion is not definitive because "each may shift its place, rising or falling gradually along the series of vertebrae." Owen wrote that *Esox* and *Thymallus* have epineurals and epicentrals, *Cyprinus* has epineurals and epipleurals, *Salmo* has epineurals together with "gristly representatives of epipleurals," whereas *Perca* and *Gadus* have only epicentrals. We regard the first of those three statements as partially true (of *Thymallus; Esox* has ossified epineurals and epipleurals), the second (*Cyprinus*) as true, the third (*Salmo*) as false (*Salmo* has epineurals and "gristly" (cartilaginous) epicentrals, not epipleurals), and the fourth (*Perca, Gadus*) as a pointer to one of the conclusions of this paper. We both initially thought Owen right, but we now think him wrong; the intermuscular bones of *Perca* (and other percomorphs) and *Gadus* (and other paracanthopterygians) are epineurals.

When all three series of intermuscular bones are present, as in *Clupea,* discriminating them is easy, and the horizontal septum (containing the epicentrals) is a useful criterion. When one (or two) series is missing and the remaining series modified, naming becomes problematic; we postpone the question of defining criteria of the different series until the concluding section of the paper, after their modifications have been discussed.

Several other names have been applied to series of intermuscular bones in teleosts. In English, these include epimerals, hypomerals, and dorsal ribs. Monod (1963:271) gave a partial synonymy of these terms and others in French and German.

The epipleural series normally is easily distinguished from epicentrals and epineurals by an ontogenetic criterion. Whereas epineurals and epicentrals develop in a rostrocaudal gradient, first appearing and being most strongly developed on the anterior vertebrae, epipleurals develop both rostrally and caudally from about the level of the first caudal vertebra. The distribution of epipleurals in *Polymixia* (Figure 1) and in many lower teleosts (Tables 3–5) illustrates the restriction of the series to the middle of the body. Inference of rostral and caudal development is borne out by the developmental series of *Albula* in Table 3 and (to a lesser extent) by the series of *Polymixia* in Table 1.

One principal aim of this paper, implied by our epigraph, is to expand the concept of intermusculars to include ligaments as well as bones. We have found that teleostean intermuscular bones almost invariably are associated with ligaments (e.g., Figure 1, Plate 1A). That association takes three forms. First, each intermuscular bone is generally an ossification within a ligament (e.g., Plate 2H), and intermuscular bones frequently are joined to the axial skeleton proximally by ligament (especially toward the caudal end of the series) and may be continued distally by ligament. Second, in a given teleostean fish each series of intermuscular bones (epineural, epicentral, epipleural) often is part of a more extensive series of ligaments;

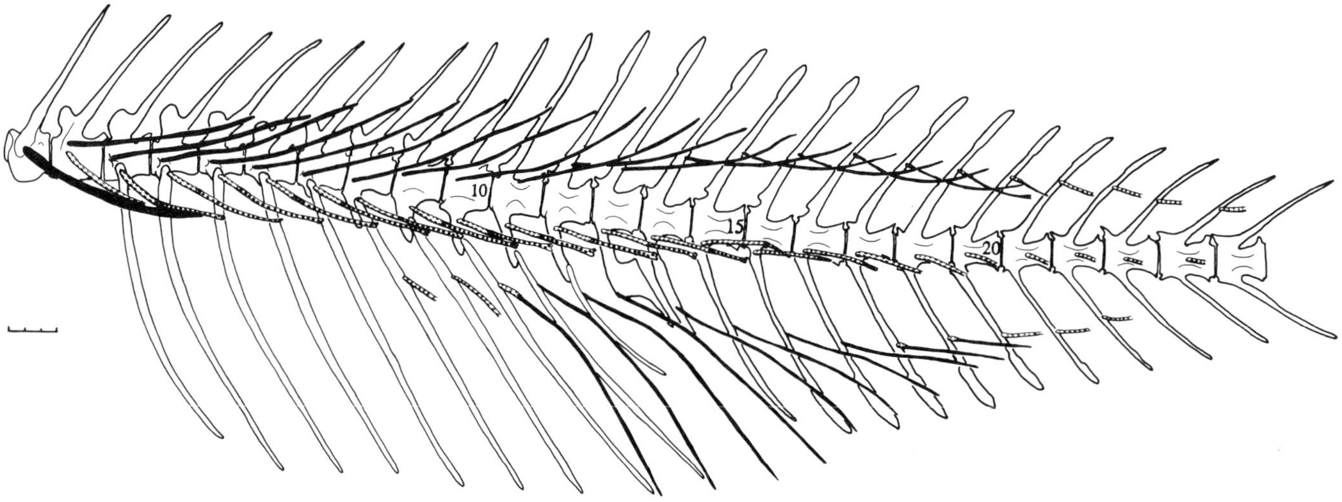

FIGURE 1.—Intermuscular system of bones and ligaments in adult *Polymixia lowei* Günther, based on BMNH 1987.12.7.1 (128 mm SL). The intermuscular bones are in solid black, the ligaments are shaded, and the epicentral cartilages are indicated by dots. (Scale bar in mm.)

the epineural, epicentral, and epipleural series of bones may be continued both rostrally and caudally by a series of ligaments (e.g., Plate 1A). Third, where a teleost appears to lack one of the three series of intermuscular bones, that series (most commonly the epicentral series) may be represented by ligaments alone (e.g., Plate 1B,C, Plate 2C,E,F). These three types of association between intermuscular bones and ligaments may be summarized as (1) ontogenetic or anatomical continuity; (2) serial homology; or (3) homology. Intermuscular ligaments occasionally are indicated in illustrations (e.g., Goodrich, 1909, fig. 336; Kishinouye, 1923, pls. 13–15) or mentioned in descriptions (e.g., Phillips, 1942:489; Nursall, 1956:139; Taverne, 1978:209; Fink and Weitzman, 1982:61), but we found almost no account of them in the literature, even in the classics of detailed anatomy (e.g., Allis, 1903, 1909). In fact, the one account we found (through the help of Mark Westneat, FMNH; also Westneat et al., 1993) is in Japanese (Kafuku, 1950) and deals only with ligaments in the horizontal septum, which contains the epicentral series in lower teleosts and the series of bones generally called epipleurals in acanthomorphs.

Kafuku studied the structure of the horizontal septum in representatives of 40 families of teleosts, including an elopomorph (*Anguilla*), three clupeomorphs (*Clupea, Sardinella, Clupanodon*), several ostariophysans (*Chanos*, three cyprinids, a cobitid) and other lower euteleosts (*Oncorhynchus, Plecoglossus*), an aulopiform (*Saurida*), and a range of acanthomorphs, including paracanthopts (*Gadus, Pterophryne*), smegmamorphs (a mugilid, an exocoetid, a scombresocid), scorpaeniforms (representatives of four families), perciforms (representatives of about 20 families), a pleuronectiform, and three tetraodontiforms. In all of these fishes he found that each vertebra gives origin to two tendons that lie in the horizontal septum, and he named them the anterior oblique and posterior oblique tendons (AOT, POT). Each AOT runs from the anteroventral part of the vertebral centrum out to the superficial lateralis muscle; according to Kafuku (1950:95) it often is cartilaginous distally, and sometimes it is called a "dorsal rib," "intermuscular bone," or "accessory rib." Kafuku found that each POT originates on the posteroventral part of the vertebral centrum and also runs out to the superficial lateralis muscle, but the POT attaching to that muscle at the same point as a given AOT may come from the centrum located two (*Scomber*), three (*Cyprinus*), or up to eight (*Katsuwonus, Thunnus*) vertebrae posterior to that on which the given AOT originates; in general, AOTs extend directly posterolaterally to the superficial muscle, and POTs run more obliquely anterolaterally, penetrating through the AOTs on their way so that AOTs and POTs form an interconnected sliding latticework. That latticework is particularly well developed in scombroids (Westneat et al., 1993) and is beautifully illustrated in pls. 13–15 of Kishinouye (1923). The number of POTs crossing a given AOT indicates the number of vertebrae separating the origin of that AOT from the origin of the POT inserting at its tip.

Kafuku (1950) made his observations of the ligaments in the horizontal septum on parboiled formalin-fixed specimens. He gave no sizes for his specimens, but they were presumably large fishes. Westneat et al. (1993) also used large, dissected fishes. We worked mainly with cleared-and-stained specimens, most of them counterstained for cartilage and bone by the method of Dingerkus and Uhler (1977). In such specimens, we found POTs to be much more spotty in their distribution than the ubiquity documented by Kafuku (1950). We also found that when POTs occur, their distribution in a fish is more complex than the uniformity throughout the vertebral column implied by Kafuku's work. POTs certainly are easier to find in larger specimens than in the small individuals usually selected for clearing and staining, and differences between our observations and those of Kafuku and Westneat et al. may in part be due to different methods of investigation. However, in general we find that the system of intermuscular bones and ligaments is fully developed in quite small individuals of the species where we have studied ontogenetic variation or size ranges (examples may be found in Tables 1, 3, 8). Furthermore, we know of no instance where study of large specimens is necessary in order to find an intermuscular bone. The only probable exception is the development of cartilage rods within epicentral ligaments; we have found well-developed rods in salmonoids and osmeroids at 40–50 mm SL, but in *Polymixia* the rods do not become visible until about 100 mm SL, and they may develop in larger individuals of other taxa.

On nomenclature, Kafuku (1950) technically was correct to call the AOTs and POTs tendons because they connect muscle (superficial lateralis) and bone (vertebra). But when an intermuscular bone or cartilage lies in one of these tendons and is not directly attached to the axial skeleton, exactitude demands that one names the proximal portion (between the intermuscular bone or cartilage and the centrum; i.e., connecting bone or cartilage to bone) a ligament and the distal portion (connecting muscle to intermuscular bone or cartilage) a tendon. We avoid this and other problems of terminology by calling them all ligaments, whether or not an intermuscular bone or cartilage is involved. Kafuku (1950) dealt only with the horizontal septum, and in that septum intermuscular bones or cartilages develop only in his AOTs. Ligaments with a broadly similar posterolateral orientation to the AOTs may occur above and below the horizontal septum, in series with or including epineural and epipleural bones. We call these epineural and epipleural ligaments, and we call the AOTs epicentral ligaments. We refer to the proximal attachment of an intermuscular ligament, on the axial skeleton, as its origin, and we call its distal attachment, if any, its insertion. We comment on the POTs in our descriptive sections, but they never contain bone or cartilage and are not part of the intermuscular skeleton on which we concentrate.

As noted above, our material is mainly cleared-and-stained specimens and, where possible, is counterstained for cartilage and bone by the method of Dingerkus and Uhler (1977). We worked primarily by surveying the BMNH and USNM collections, borrowing material from other institutions where necessary. (Institutional abbreviations follow Leviton et al. (1985).) Rather than give a full list of material examined (which would run to many pages), in Appendices 1 and 2 we give alphabetical and systematic lists of all genera cited in the text and in Tables 1–5, 7, and 8. We examined specimens of many other genera beyond those listed in the Appendices. In addition to cleared-and-stained specimens, we studied some that were alizarin-stained without clearing before dissection, some dissected spirit specimens, and many dried skeletal preparations and radiographs of larger specimens. In Tables 3–5, 7, and 8, summarizing the intermusculars and associated structures in a wide variety of teleosts, we generally record structures on the left side of the specimen. Where there is noticeable variation between the two sides of a specimen, we record the state that appears normal, especially when we had access to more than one individual.

Although there can be little argument about the distribution of intermuscular bones in cleared-and-stained specimens, we cannot pretend that observing and recording intermuscular ligaments is an entirely objective procedure, particularly toward the caudal end of a series of ligaments, where they decrease in size and coherence. We have found that the condition of the specimen is important (how well digested and/or bleached), as is the lighting and the quality of the optical equipment. Incident light, in addition to or instead of transmitted light, often is useful in resolving the ligaments (e.g., Plate 2A–C), but too much reliance on incident light may overemphasize connective tissue sheets that do not deserve the term ligament. Transferring cleared-and-stained specimens to alcohol may render the ligaments more opaque, and so more visible. We have found that Leitz stereomicroscopes give the best resolution and may be essential in resolving details of ligaments in small specimens. In cases of doubt, we have changed illumination and observer until agreement was reached. Nevertheless, the observations of ligaments recorded in our tables include an unknown quantity of subjectivity, and we will be glad to see all or any of them checked by others. Our decisions, also recorded in the tables, on whether the pleural ribs of a specimen are preformed in cartilage or are membrane bone is based on the distinction between a rib tip that is an open, cartilage-capped cylinder and one that ends in a bony point. That distinction may be hard to draw in ribs that are very slender, or poorly stained, or decalcified, and there is an element of subjectivity.

A NOTE ON THE TABLES

In Tables 1–3, 7, and 8 we record features of the intermuscular bones and ligaments, and other aspects of the vertebral column, in about 125 genera of teleosts from over 100 families. The entries in those tables are, in effect, maps of the vertebral column or a form of illustration in which symbols stand in for objects. As with maps, some effort is necessary to learn the conventions. It will help readers who intend to study or use the tables, in conjunction with either our text or their own specimens, to make a photocopy of the "Key to the Tables" (p. 50), a photocopy of Table 3, which is printed as a foldout, and photocopies of Tables 4, 5, 7, and 8 which may be cut and pasted to make the equivalent of foldouts of those tables.

ACKNOWLEDGMENTS

For the loan of specimens, we are most grateful to William D. Anderson Jr., Douglas P. Begle, Edward J. Crossman, M. Norma Feinberg, Anthony C. Gill, Karsten E. Hartel, Dannie A. Hensley, Robert K. Johnson, Gareth Nelson, Muneo Okiyama, James W. Orr, John R. Paxton, William J. Richards, Richard H. Rosenblatt, Darrell J. Siebert, Melanie L.J. Stiassny, Kenneth J. Sulak, Yoshiaki Tominaga, James C. Tyler, and Richard Winterbottom. For unpublished information we thank Lance Grande, Tony Harold, Gordon J. Howes, P. Humphry Greenwood, John E. Olney, Lynne R. Parenti, Darrell J. Siebert, and Mark W. Westneat. For help in translating Japanese we are indebted to Louise and Seiji Arimatsu. We thank John Harshbarger for preparing histological sections. For reviewing the manuscript and providing many constructive suggestions for its improvement we are particularly grateful to George V. Lauder, Gareth Nelson, Victor G. Springer, Melanie J. Stiassny, and Richard L. Zusi. And for technical and logistic support of many kinds we are most obliged to Carole C. Baldwin.

Intermusculars in *Polymixia*

The description that follows is based on four counterstained specimens of *P. lowei* Günther: MCZ 64773, 12 mm SL; MCZ 95714, 28 mm SL; USNM 308378, 72 mm SL; and BMNH 1987.12.7.1, 128 mm SL; features of osteology also were checked in dried skeletons of *P. japonica* Günther and *P. nobilis* Lowe. *Polymixia* (Figure 1, Plate 2A–D) does not have two series of intermusculars (Starks, 1904a; Patterson, 1964; Zehren, 1979), rather it has three: epineurals above the horizontal septum, epicentrals in that septum, and epipleurals below it. Ligaments occur in association with all three series.

In the adult, the entire upper, or epineural, series of bones and ligaments extends over about 24 vertebrae, from the first vertebra (V1) back to about the twelfth caudal (PU5). The epineural bone on V1 differs from its successors in being larger and in lying in the horizontal septum, so that it is in series positionally with the epicentral ligaments behind it; we anticipate our discussion of the problem of its homology (p. 33) in describing it as an epineural. The first epineural is a stout bone, originating in a socket at the base of the (autogenous) first neural arch, at a level slightly ventral to the articulation of the epineural on the second neural arch (Rosen, 1985, fig. 18). The bone extends posterolaterally in the horizontal septum and ends immediately beneath the lateral line nerve, at the level of V4. On V2 the epineural articulates at the base of the (fused) neural arch, but on V3–10 the epineurals originate on the centrum. On V3–8 the attachment is on the upper edge of the

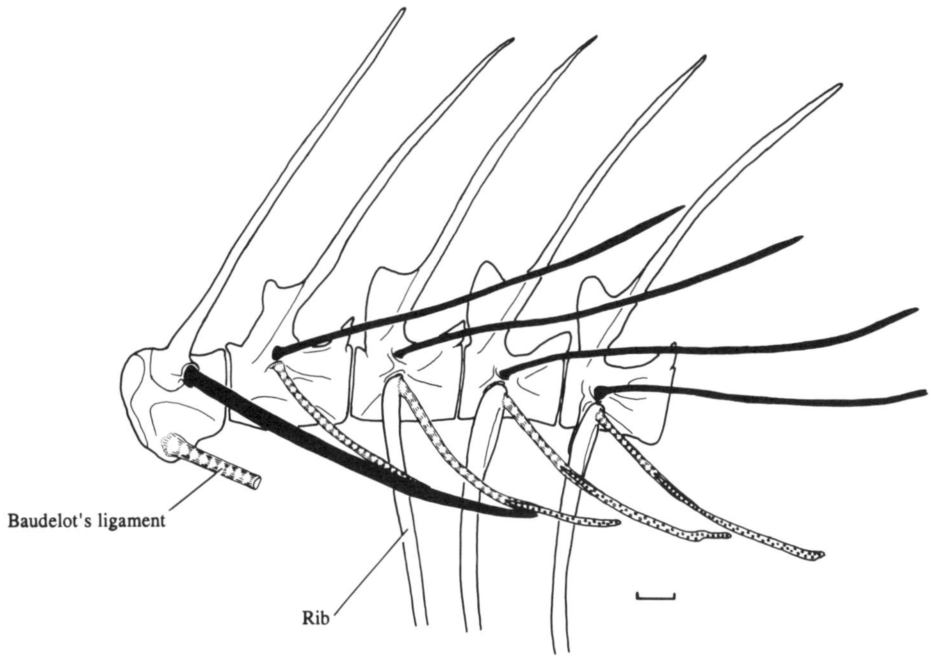

FIGURE 2.—First five vertebrae and attached intermusculars in *Polymixia lowei,* based on BMNH 1987.12.7.1 (128 mm SL) and supplemented by USNM 308378 (72 mm SL). (Scale bar (1 mm) refers to the larger specimen.)

parapophysis, just above the head of the rib, and on the succeeding one or two vertebrae (V9–10) it is at or just above the midpoint of the anterior margin of the centrum. On V11 the epineural originates on the base of the neural arch, at about the same level as on V2, and on more posterior vertebrae the point of attachment of the epineural rises successively higher on the neural arch and spine. The last five to seven epineural bones are attached to the axial skeleton by ligaments that are progressively longer on more posterior vertebrae, and in the adult there are ligaments, with no included bone, on about four more neural spines, so that the entire series extends back to about PU5.

The lowermost, or epipleural, series of bones in *Polymixia* extends over some 10 vertebrae, from about V9 back to V19 (the seventh caudal). The first and the last one or two epipleural bones have no direct articulation with the axial skeleton but are attached by ligaments, and in front of the first ossified epipleural and behind the last there are from one to three ligaments with no included bone (Plate 2G). The remaining epipleurals attach near the heads of the last two or three pleural ribs and at successively lower points on the first five or six haemal spines (Figure 1).

Between the upper and lower intermuscular series of *Polymixia* there is a middle series, the epicentrals. As noted above, the first epineural lies in the horizontal septum. Behind and in series with this stout bone in the horizontal septum there is a row of epicentral ligaments (Plate 2B–D), one in each myocomma, extending from V2 back to the posterior caudal vertebrae (to about PU5 in the adult). Our epicentral ligaments are the anterior oblique tendons (AOTs) of Kafuku's (1950) account of the teleostean horizontal septum. In *Polymixia* they are stout, cylindrical ligaments, whereas the posterior oblique tendons (POTs of Kafuku, 1950) are very feeble, each consisting of several strands. As shown in Figure 3, three POTs cross each epicentral ligament, and the POT inserting with a given epicentral ligament originates on the vertebra four behind that on which the epicentral ligament originates. The first epicentral ligament (Figure 2) originates immediately below the attachment of the epineural on V2. On V3 and succeeding abdominal vertebrae the ligament originates on the head of the rib, immediately below the origin of the epineural on the parapophysis. On anterior caudal vertebrae the ligament originates at the base of the haemal spine, and from about the sixth caudal vertebra onward it originates on the centrum. In the distal part of the second to sixteenth of these ligaments (V3–17) in our 128 mm SL specimen there is a series of cartilaginous rods (Plate 2D). We identify the tissue of these rods as cartilage by histology (in thin section and in squash preparations; it is cellular with much extracellular matrix), by comparing its reaction with alcian blue (deeply stained) with that of undoubted cartilage in the same specimen, and by comparison with similar rods in *Salmo* (Plate 1D and below), both in thin section and in double-stained specimens, where cartilage previously has been verified histologically (Emelianov, 1935). Like the epineural bone on V1, each of these cartilaginous rods ends distally beneath the lateral line nerve. They are most strongly chondrified distally, where their tips are often bifid, and they grow progressively shorter posteriorly.

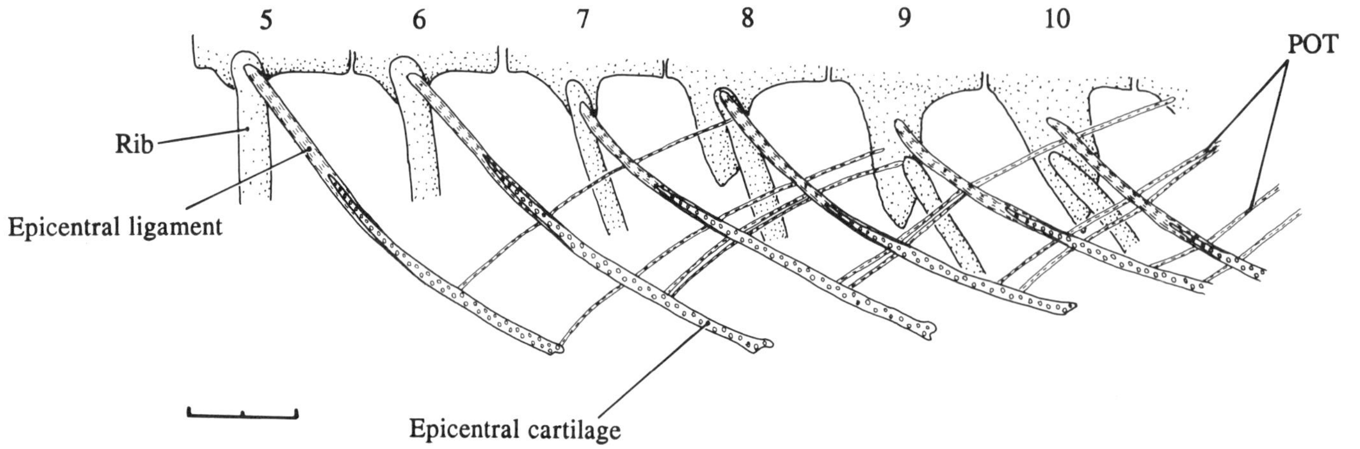

FIGURE 3.—Epicentral ligaments and posterior oblique ligaments (POTs) in horizontal septum in abdominal region of *Polymixia lowei*. The sketch, in dorsolateral view, shows the parapophyses and rib heads of V5–10 in BMNH 1987.12.7.1, with the epicentrals and POTs attaching to them. The cartilage in epicentral ligaments is shown by open circles. The vertebrae are numbered. (Scale bar in mm.)

Associated with the epicentral ligaments of *Polymixia* are the posterior oblique tendons (POT) of Kafuku (1950). The POTs lie in the horizontal septum and are clearly visible in our 128 mm specimen (Figure 3). Working back from the head, the first two POTs insert on the epicentral ligament of V5; one POT inserts about two-thirds of the way along that epicentral and another inserts at its tip where the epicentral ends in the lateralis muscle. These two POTs originate respectively on the head of the rib of V8 and the parapophysis of V9. Another slender POT originates on the head of the rib of V9 and inserts on the epicentral of V6. V10 gives origin to no POT. A POT originating on the parapophysis of V11 ends on the epicentral of V7. The POT originating on the parapophysis of V12 (the last abdominal) inserts on the epicentral ligament of V8, and the POT originating on the haemal arch of V13 inserts on the epicentral of V9. This pattern is repeated back into the caudal region, with the POT inserting on a given epicentral ligament originating on the vertebra four behind. From V11 back, the point of insertion of the POT is always at the tip of the epicentral ligament rather than some distance from its tip as it may be more anteriorly. Passing back along the column, the point of origin of the POTs rises from the haemal spine to the anteroventral margin of the centrum, and at V19–20 it shifts to the posterior part of the preceding centrum, so that V19 carries two POTs, one from the anteroventral and one from the posteroventral part of the centrum. From this point back to the last epicentral ligament (on V24) epicentrals and POTs show the pattern described by Kafuku (1950) as general in teleosts, with the epicentral originating at the front of a centrum and the POT at the rear. This pattern exists only on about five centra in *Polymixia*. The last visible POT originates on V25 (PU4). The POTs of *Polymixia* are very feeble in comparison with the epicentral ligaments. The epicentrals are stout cylinders, lying on the dorsal surface of the horizontal septum, whereas the POTs are feeble, dorsoventrally compressed, strap-like ligaments lying within the horizontal septum. The POTs pass ventral to the epicentral ligaments (although attached to the epicentral by connective tissue) and do not pass through or penetrate the epicentrals, which is the situation described by Kafuku (1950) in other teleosts. In the middle of the trunk, at approximately the first caudal vertebra, POTs are about half the size of the epicentral ligament originating on the same centrum. As they pass forward, these POTs break up into two or three slender strap-like ligaments. Posteriorly, near the end of the series, each POT divides into five or six separate strap-like ligaments almost immediately after its origin from the centrum. Anteriorly, at the front of the series, where the epicentral ligaments are more robust than posteriorly, the contrast between their size and the diminutive POTs is even more obvious (Figure 3). The anterior POTs divide into two or three branches, like those in the middle of the series. In our 72 mm SL specimen, POTs show much the same pattern as in the 128 mm SL specimen. We have not succeeded in seeing POTs in our two smaller specimens of *Polymixia* (12 and 28 mm SL).

The ribs of *Polymixia* extend from V3 to V12. All are ossified in cartilage.

Table 1 introduces our method of recording intermuscular bones and ligaments in teleosts and shows the conditions in four specimens of *Polymixia* ranging from 11 to 128 mm SL.

Intermusculars in Nonacanthomorph Teleosts

Polymixia has three sets of intermusculars that match Owen's (1866) topographic criteria for epineurals, epicentrals, and epipleurals. The situation in *Polymixia* is relatively uncomplicated because all the intermuscular bones are simple;

none shows the proximal or distal forking or branching that occurs in many lower teleosts, which may render their topography and relationships more difficult to decipher. In this section we first describe the intermusculars in the gonorynchiform *Gonorynchus,* as an example of a nonacanthomorph with the most extreme complications of that sort. (In keeping with our epigraph, the intricacies of the intermuscular bones and ligaments in *Gonorynchus* are not necessarily mere worthless detail for they may be phylogenetically informative.) We then review the structure and distribution of the three series of intermusculars and some associated structures in nonacanthomorph teleosts, and finally we comment on the significance of variation in the intermusculars within and between some major groups of nonacanthomorphs.

Gonorhynchus

The description that follows is based on SIO 70-275, 146 mm SL (Figures 4, 6); we also include a drawing of a smaller double-stained specimen (Figure 5) made by one of us some years ago. Monod (1963) gave a full account of intermuscular bones in larger specimens (~300 mm SL). The situation in *Gonorynchus* is so complex that in the description we do not try to take account of variation among individuals (which certainly occurs, and may include interspecific variation, cf. Howes, 1985:280). We compare our observations with Monod's (1963) at the end of the description.

The epineural series in *Gonorynchus* ostensibly extends from the fourth vertebra (V4) back to about V47 (PU9), but as we will show, epineurals extend anterior to V4. In our 146 mm specimen there are ossified epineurals that attach to neural arches of V5–40 and unattached epineural bones on V41–47. These epineural bones are all associated with or ossified within ligaments, and those on V5–40 are attached to the anterior part of the neural arch by ligament (Figures 5, 6). Those on V5–35 are forked proximally (Figures 4–6), with a primary anteromedial branch attaching to the neural arch and a secondary anteroventral branch passing toward the transverse process (fused parapophysis) of the next vertebra in front. Passing back along the series, the point at which the epineural bifurcates becomes gradually more remote from its attachment to the neural arch, and at V36, in the region of the origin of the dorsal fin, the anteromedial branch ceases to join the distal part and anteroventral branch; beyond this point there are five or six short, simple epineurals attached to the neural arch by ligament, and lateral to them is a corresponding series of long bones that represent the distal portions and anteroventral branches of the posterior epineurals. This series of bones extends back to V47. At the anterior end of the epineural series there is a similar separation of the epineurals into two parts. On V4 there is a short, simple epineural bone (En 4, Figure 4), which differs from its successors in attaching to the anterodorsal part of the centrum, not to the neural arch, and in articulating directly with the vertebra, with no intervening ligament. If one traces the distal parts of the epineurals forward in sequence, it is evident that this short and simple epineural on V4 is merely the isolated anteromedial branch of an epineural, and that the distal part and anteroventral branch of this epineural exists as a long, separate bone (V_0 4, Figures 4, 5), extending down (like its successors) toward the region of the parapophysis of the next vertebra in front, V3. That vertebra bears the first rib, which is enlarged and articulates with a short transverse process. The tip of the epineural branch of V4 is directed toward the head of the rib rather than toward the transverse process. On V3 there is a ligament attaching to the centrum (En 3, Figure 4), a little lower than the attachment of the epineural bone on V4, and on the right side of our 146 mm specimen there is a diminutive epineural bone within the ligament. No epineural bone or ligament attaches to V1 or V2. However, the epineurals of V1 and V2, which appear to be missing when only vertebrae are considered, are represented by their distal parts and anteroventral branches, each directed proximally toward the axial skeleton of the segment in front. But on V1–3 there is an additional complication in that the anterior ends of these three epineural branches are fused or coossified with the anterior members of the epicentral series; these epineural branches are described with the epicentral series, below.

The epicentral series of *Gonorynchus* extends from the cranium back to about V50, having about the same posterior extent as the epineurals. In our 146 mm SL specimen there are epicentral bones back to V40 and epicentral ligaments thereafter. Epicentral bones and ligaments attach to the tips of the transverse processes of abdominal vertebrae, to the base of the haemal arch of the first caudal vertebra (V43), and to the centrum on posterior caudal vertebrae. They lie in the horizontal septum, and all epicentral bones are simple, unbifurcated rods except for those on the occiput and V1–2, which are modified and associated with the anteroventral branches of the first few epineurals. On V4 the epicentral is a substantial bone (Ec 4, Figures 4, 5) that attaches to the anterior margin of the large transverse process. On V3, which carries the enlarged first rib, there is apparently no epicentral. On V2 there is no rib, but there is a bifid transverse process. Attached by ligament to the posterior limb of the transverse process is a substantial intermuscular bone (Ec 2 + V_0 3, Figures 4, 5). That bone bifurcates almost immediately into a posterodorsal branch and a posterolateral branch. The posterodorsal branch is in series posteriorly with the epineural bones, and it is the detached distal part and anteroventral branch of the epineural of V3. The posterolateral branch is in series posteriorly with the epicentrals, and it is the epicentral of V2. V1 also has a bifid transverse process, like V2, and attached by ligament to the two limbs of the process are two large intermuscular bones. The proximal end of both may be bifid or trifid, and the bones are closely associated just beyond their attachment. The more posterior of the two bones (Ec 1 + V_0 2, Figure 5) splits into two branches, directed respectively posterodorsally and posterolaterally and agree with the epineural and epicentral branches of the intermuscular on V2 (this epineural branch on V1 represents the epineural of V2). The more anterior intermuscu-

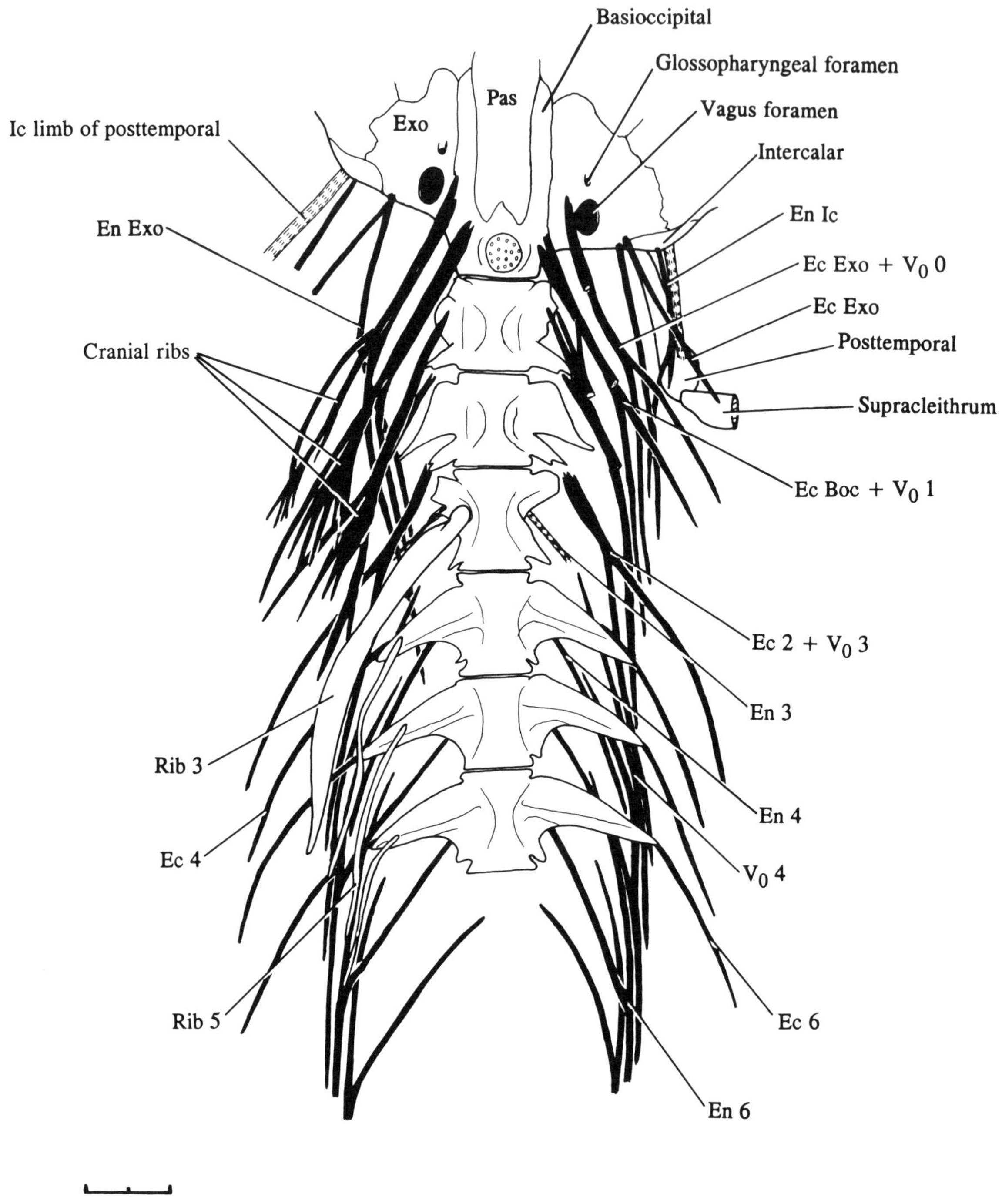

FIGURE 4.—*Gonorynchus* ?*gonorynchus* (L.), SIO 70-275, 146 mm SL. Ventral view of the first six vertebrae and the rear of the braincase, with the intermuscular bones in solid black. The ribs are removed on the left side (right side of figure). The left "cranial ribs" on the occiput and V1, and the epicentral (lower) branch of left compound intermuscular on V1 are drawn as if they were cut through near their origin. The left posttemporal and supracleithrum are included, with the supracleithrum drawn as if it was cut through. The only ligaments shown are those representing the intercalar limb of the posttemporal and the epineural ligament of V3, which includes a diminutive bone on the right side. (Scale bar in mm.) (Abbreviations: Boc, basioccipital; Ec, epicentral; En, epineural; Exo, exoccipital; Ic, intercalar; Pas, parasphenoid; V_0, intermuscular representing unattached anteroventral limb and posterior body of an epineural. Plus signs between symbols indicate fusion. Intermusculars and ribs are numbered or labeled according to the vertebra (number) or bone (abbreviation) with which they are associated.)

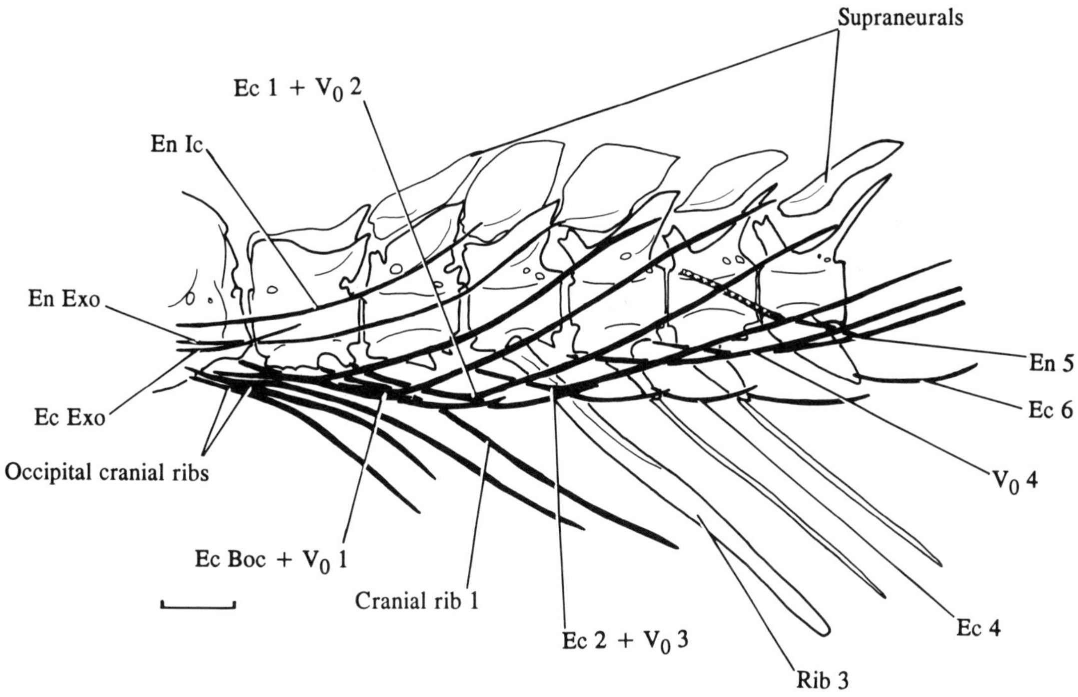

FIGURE 5.—*Gonorynchus ?gonorynchus*. Anterior vertebrae and intermuscular bones in left dorsolateral view of RUSI 15048, 95 mm SL. The intermuscular bones are in solid black. The only ligament shown is that attaching the fifth epineural to the neural arch. (Scale bar 1 mm.) (Abbreviations: Boc, basioccipital; Ec, epicentral; En, epineural; Exo, exoccipital; Ic, intercalar; V_0, intermuscular representing unattached anteroventral limb and posterior body of an epineural. Plus signs between symbols indicate fusion. Intermusculars and ribs are numbered or labeled according to the vertebra (number) or bone (abbreviation) with which they are associated.)

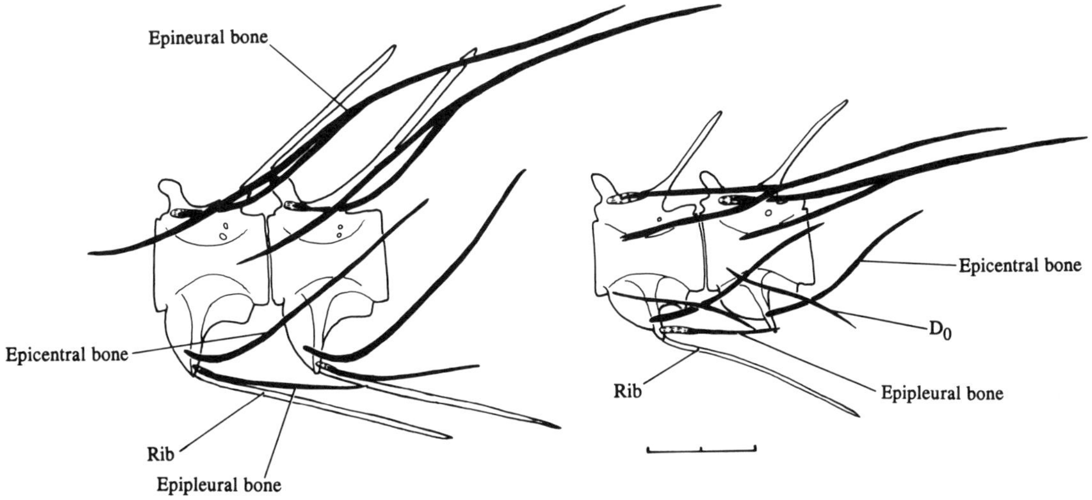

FIGURE 6.—*Gonorynchus ?gonorynchus*, same specimen as Figure 4. Lateral view of vertebrae 25 and 26 (left) and of vertebrae 34 and 35 (right), with attached intermuscular bones and ligaments. The intermuscular bones are in solid black. (Scale bar in mm.) (Notation D_0 (as in Table 2) denotes unattached bones that are homologous with anterodorsal branch of an epipleural.)

lar on V1 is directed posteroventrally in the peritoneal membrane, so that it lies in the plane of the ribs distally, where it breaks up into a brush-like termination, with about seven branches passing out into the musculature; we interpret this bone as the serial homologue of the two "cranial ribs" (below). Anterior to these complex intermusculars on V1 there are seven more intermusculars, which insert on the cranium. These include the "cranial ribs" of previous descriptions of *Gonorynchus* (Chabanaud, 1931; Monod, 1963). The two most posterior of these bones insert close together on the basioccipital, just anterior to the lower margin of the occipital condyle. The more dorsal of the two (Ec Boc + V_0 1, Figures 4, 5) bifurcates into two branches, one posterodorsal and one posterolateral, respectively matching the epineural and epicentral branches of the succeeding bones (so that the epineural branch represents the epineural of V1). The more ventral bone is directed posteroventrally, breaks up into a brush-like termination in the peritoneal membrane, and is a cranial rib, matching the anterior intermuscular on V1 in orientation and configuration. The next two intermusculars insert close together on the exoccipital, immediately beneath the vagus foramen. They show the same pattern as the two bones on the basioccipital, with the more dorsal bone (Ec Exo + V_0 0, Figure 4) bifurcating into epineural and epicentral branches and the more ventral ending in a bony brush in the peritoneal membrane in series with those of the two succeeding cranial ribs. Passing forward on the cranium, the next two intermusculars (Ec Exo, En Exo, Figures 4, 5) insert close together on the exoccipital just above and lateral to the vagus foramen. They are smaller than the bones behind them on the cranium, and together they match the upper of the two intermusculars on the exoccipital and on the basioccipital; the upper bone matches their epineural branch, and the lower matches their epicentral branch. The lower, epicentral, bone bifurcates and both limbs attach to the medial face of the supracleithrum. The last intermuscular on the cranium (En Ic, Figures 4, 5) is a slender bone attached to the intercalar; this intermuscular bifurcates distally, with an outer branch attaching to the medial face of the posttemporal and an inner dorsolateral branch (which bifurcates again) passing back into the musculature in series with the epineurals. Ridewood (1904:66) homologised this last intermuscular with the intercalar limb of the posttemporal, which he thought was otherwise absent in *Gonorynchus,* but we believe that the intercalar limb is represented by a stout, strap-like ligament that originates on the intercalar immediately beneath the point of origin of the epineural intermuscular and passes out to the posttemporal (Figure 4).

The epipleural series in our 146 mm SL *Gonorynchus* ostensibly extends from V11 back to V35 (the level of the pelvic-fin insertion), but, as we will show, it extends much further posteriorly. All the epipleurals to V35 are ossified, and there appears to be no rostral or caudal extension of the series in ligament. The bones are simple rods, attached by short ligaments to the tips of transverse processes (Figure 6). They are directed slightly posterodorsally, and most of them extend back in the musculature for a distance equal to about one-and-one-half vertebrae. The first few and last few bones are shorter than those in the middle of the series and are attached to the transverse process by longer ligaments. Beginning at V30, and extending posteriorly to V52, there is another series of epipleural bones that clearly mirror the detached anteroventral branches of the posterior epineural bones. These bones (D_0, Figure 6) lie unattached (free) in the lateral musculature above the epipleurals and lateral to the middle part of the epicentrals; they are oriented obliquely ventrolaterally, and, in the anterior part of the series, each begins close to the middle part of the epicentral bone of the preceding vertebra. The first few bones are short (about the length of a vertebra), but they lengthen posteriorly (up to about two vertebrae long) and become oriented more horizontally.

We have not found any posterior oblique tendons (POTs) in *Gonorynchus,* and Baudelot's ligament also is absent.

Ribs of *Gonorynchus* extend from V3 to V34. The first is enlarged and is preformed in cartilage. The remainder are slender and ossified as membrane bone.

Table 2 illustrates our method of recording intermuscular bones and ligaments as applied in *Gonorynchus.*

Monod (1963) gave an excellent account of intermuscular bones in *Gonorynchus* based on dissected specimens two or three times as large as ours (~300 mm SL). His observations match our own in most respects. Monod's specimen has 62 preural vertebrae, whereas our 146 mm specimen has 55. He used an alphabetical notation (*a*1–50, *b*1–50, etc.) for the different series, but he concluded that his *a* series comprised epineurals, his *b* series epicentrals, his *c* series ribs, and his *d* and *e* series epipleurals. Monod's *e* series extends from V33 into the caudal region, and posteriorly from V49 onward each *e* bone is fused with the tip of an epipleural (Monod's *d*), so that his *e* bones correspond to the anterodorsal branch of the epipleural (D_0 in Table 2 and Figure 6). In our smaller specimen (146 mm SL), principal epipleurals (attached to the axial skeleton) extend only to V35, overlap the *e* series (beginning at V30) for only a few segments, and are nowhere fused with them. Fusion between the two may occur in larger individuals, as the series of principal epipleurals extends to more posterior vertebrae. Monod found that the principal (anteromedial) and distal (anteroventral branch + posterior body of the bone; notation V_0 in Table 2 and Figure 6) branches of epineurals did not separate until about V51, and that the medial branch did not disappear until V54 (cf. V36 and V41 in our specimen). On the back of the cranium and the first few vertebrae, where the intermusculars of *Gonorynchus* are most intricate, Monod's observations and interpretations agree with our own in most details.

Monod included three "cranial ribs" in his *c* series, two originating on the cranium and one on V1. We can find no other interpretation for these three bones, which (as noted above) appear to lie in the peritoneal membrane. We have not found cranial ribs in any other teleost (the supposed "cranial ribs" of *Chanos* and other gonorynchiforms are epicentrals, Table 4) and infer that they are autapomorphous for gonorynchids. We do not believe that they are homologous with "true" ribs, and we use a different symbol for them in Table 2.

Monod failed to notice that the anteroventral (secondary) limb of each epineural is directed toward the transverse process of the vertebra in front of that to which the primary (anteromedial) limb attaches. He also referred to the primary limb, attaching to the neural arch, as the accessory branch, and he took the anteroventral limb and the posterior part of each epineural to represent the principal part of the bone. He found no intermusculars on V2 and sought to explain this by a phylogenetic shift of the intermusculars from the posterior part of one centrum to the anterior part of its successor. Intermusculars exist on V2 in our specimens, so no such explanation is necessary. The absence of an epicentral on V3 seems more to demand explanation, and we return to it below, in the section on ostariophysans.

Gonorynchus has an intermuscular system more intricate and challenging than that of any other teleost in our sample.

THE INTERMUSCULAR SERIES AND ASSOCIATED STRUCTURES IN OTHER NONACANTHOMORPH TELEOSTS

Tables 3–5 record the distribution and form of intermuscular bones and ligaments in a range of nonacanthomorphs. In this section we comment on the three series of intermusculars in those fishes and on their POTs, ribs, and certain other features recorded in the tables.

EPINEURALS.—The primitive teleostean condition with regard to intermuscular bones is to have epineurals alone (Patterson and Rosen, 1977:129; Schaeffer and Patterson, 1984:61). Actinopterygian epineurals are primitively developed as short, broad outgrowths of the anterior neural arches that are preformed in cartilage and ossify perichondrally (in cartilage bone) so that in fossil actinopterygians they are hollow and open at the tip, which was cartilage-capped in life. Short epineural processes of this type occur on the first few vertebrae in Devonian lungfishes (Rosen et al., 1981, fig. 54) and palaeoniscoids (Gardiner, 1984, figs. 121–124) and may well be primitive for osteichthyans. Short, cartilage-tipped epineural processes occur on about the first 20 vertebrae in the Jurassic neopterygian *Hulettia* (Schaeffer and Patterson, 1984, figs. 15, 17C). It is synapomorphous for teleosts to have much longer epineurals, equal in length to several vertebrae (e.g., the early Jurassic *Pholidophorus bechei*, Schaeffer and Patterson, 1984, fig. 17D). These long epineurals were still hollow and cartilage-tipped in *Pholidophorus*, a condition that we observed in *P. bechei* (the type species), the early Jurassic *P. germanicus*, and two late Jurassic species, "*P.*" *macrocephalus* and the species described by Patterson (1975) as "the Callovian *Pholidophorus* sp." The cartilaginous anlagen of the epineurals evidently have regressed in all Recent teleosts, as have those of some or all the pleural ribs in many teleosts (Emelianov, 1935; Tables 2–5, 7, 8). We have seen no sign of cartilage in epineurals of any Recent teleost (including the developmental series of some taxa), and Emelianov (1935) found no cartilage in epineural bones of the teleosts that he investigated, which included *Salmo* and a clupeoid. It is therefore synapomorphous for Recent teleosts to have solid, membrane bone epineurals, and this derived condition also characterizes some Jurassic stem-group teleosts. We observed solid, membrane bone epineurals, without cartilage cores or tips, in the late Jurassic "leptolepids" *Ascalabos, Tharsis,* and the species described by Patterson (1975) as "the Callovian *Leptolepis* sp.," as well as in the late Jurassic ichthyodectiform *Allothrissops* and the late Jurassic elopocephalan *Anaethalion.* However, in the early Jurassic *Leptolepis coryphaenoides* (type species of the genus), the epineurals are hollow and evidently had a cartilage core in life, but their tips are fully ossified, so that there was no cartilage cap. In the early Jurassic *Proleptolepis,* the epineurals seem to show the same condition: they are hollow, but the tips of at least the most posterior ones (our material is incomplete) are fully ossified. Solid, membrane bone epineurals, therefore, seem to characterize the node above *Leptolepis coryphaenoides* in the cladogram in Patterson and Rosen (1977, fig. 54), or ichthyodectiforms, *Ascalabos, Tharsis,* and all higher teleosts, and epineurals with a cartilage core but without a cartilage tip characterize a node above *Pholidophorus bechei* and below *Leptolepis coryphaenoides* in that cladogram. Those two taxa are separated by *Pholidolepis dorsetensis* in the cladogram. We examined the available material of that species, and although we can confirm that the epineurals had a wide cartilaginous core, we could see no well-preserved epineural tips; the lack of distal tapering of the epineurals is suggestive of the condition in *Pholidophorus* (cartilage-tipped) rather than that in *Leptolepis* (solid tips).

Primitive continuity between epineurals and neural arches persists in all epineurals of *Pholidophorus bechei* and of *Leptolepis coryphaenoides.* In both genera, epineurals are confined to the abdominal region. In the derived late Jurassic "*Pholidophorus*" *macrocephalus,* only about the first eight epineurals are continuous with the neural arch, and the remainder (they extend back to about V22, and the first caudal is V27) are free, articulating with a socket on the neural arch. In the ichthyodectiforms *Thrissops* (Taverne, 1977a) and *Allothrissops,* all epineurals are continuous with the neural arch; they extend back to the third or fourth caudal vertebra. In *Tharsis,* epineurals extend back to about the first caudal vertebra, and there appear to be two or three free epineurals posteriorly, articulating with the neural spine. In *Ascalabos,* epineurals extend to about the fifth caudal vertebra, and, as in *Tharsis,* the last two or three are free and articulate with the neural spine. As in the above stem-group teleosts, primitive continuity between anterior epineurals and neural arches persists in various more-derived fossil and living teleosts (e.g., the Jurassic *Anaethalion* and Recent *Elops,* Arratia, 1987, figs. 3, 31; *Hiodon,* Taverne, 1977b, figs. 22, 23; *Pantodon,* Taverne, 1978, figs. 49–53; *Esox,* Figure 8; *Alepocephalus,* Gosline, 1969, fig. 11; *Clupea,* Figure 7A, and Ramanujam, 1929:389; stomiiforms, Weitzman, 1974, fig. 85; Fink and

Weitzman, 1982, fig. 5). In all these forms (except the osteoglossomorphs *Hiodon,* where all epineurals are fused to the neural arch (Table 3), and *Pantodon,* where all but the first are so fused (Taverne, 1978)), the more posterior epineurals articulate with the neural arch or spine, and the most posterior ones are either free or connected with the neural spine by ligament. In more-derived teleosts, anterior epineurals are separate, either articulating with the vertebra (e.g., Rosen, 1985, figs. 9, 11, 13, 15–17) or attaching to it by ligament, and posterior epineurals are free from the vertebral column and often are forked proximally (e.g., Figures 4, 5, 8, Plate 1E,F Weitzman, 1962, fig. 14; Jollie, 1962, fig. 6.40, "epimerals"), a condition that we take to be synapomorphous for elopocephalans. When epineurals are forked anteriorly (as in *Gonorynchus,* above), there is an anteromedial branch that represents the primitive connection with the neural arch and an anteroventral branch that passes forward in the myoseptum, roughly parallel to the vertebral column, into the epaxial forward flexure of the myomere (Nursall, 1956, fig. 8).

Epineurals are virtually universal in nonacanthomorph teleosts, occurring in all "leptolepids," ichthyodectiforms, osteoglossomorphs, elopomorphs, and clupeomorphs, in ostariophyans (except siluroids), in esocoids, argentinoids, osmeroids (osmeroids sensu Rosen, 1974, were said by Rosen to lack intermusculars, but all osmeroids that we have examined have epineurals), salmonoids, stomiiforms, aulopiforms, and myctophiforms (cf. Tables 3–5).

The only nonacanthomorphs lacking epineural bones among those recorded in Tables 3–5 are the southern osmeroids *Stokellia* and *Lepidogalaxias,* the salangid *Salangichthys,* and the umbrid *Novumbra* (they also are lacking in *Dallia*). In *Stokellia, Novumbra, Dallia,* and salangids, the epineural series is represented by a series of ligaments with the same distribution as epineural bones in other lower euteleosts. We found the same condition in *Retropinna,* and it may be a synapomorphy corroborating the sister-group relationship between that genus and *Stokellia* (Begle, 1991), but McDowall (1971, 1976) reported that ossified epineurals also are absent in *Aplochiton, Lovettia,* and *Prototroctes.* In *Lepidogalaxias,* there are no epineural bones or ligaments, a unique condition. In other osmeroids (*Osmerus, Hypomesus,* Table 4) and various other lower euteleosts (e.g., the stomiiform *Diplophos* and the aulopiforms *Ahliesaurus, Scopelosaurus, Scopelarchoides, Synodus,* and *Trachinocephalus,* Table 5), the series of epineural bones is extended caudally by a series of ligaments, without included bone. In the evermanellid aulopiform *Coccorella,* there is only one epineural bone (true of all evermanellids according to Johnson, 1982:55), and the epineural series is continued caudally as ligament.

Epineurals are posterodorsally inclined in the epaxial backward flexure of the myoseptum (Nursall, 1956, fig. 8), and their distal tips are normally (and primitively) all aligned at the same level. But in *Polymixia* (Figure 2), the bone that we interpret as the first epineural is out of alignment with the rest of the series and lies in the horizontal septum. In a few nonacanthomorphs, the distal part of the first one or more epineurals also is displaced ventrally relative to their successors, although, in contrast to *Polymixia,* they are always above the horizontal septum. We have observed this ventral displacement in alepocephaloids, argentinoids, several aulopiforms, and the neoscopelid *Neoscopelus* (notation b_1 or f_1 in Tables 4, 5). In alepocephaloids and argentinoids, the tips of the first three (*Glossanodon, Leptoderma, Searsia*) or four (*Argentina*) epineurals are displaced, a unique condition that we take to be synapomorphous, corroborating the sister-group relationship of these two taxa (Greenwood and Rosen, 1971; Begle, 1991). In aulopiforms, the tips of the first one to three epineurals may be displaced, and in *Neoscopelus* only that of the first is displaced (none is displaced in the other neoscopelid genera, *Scopelengys* and *Solivomer*).

In *Polymixia,* the proximal articulation of the epineurals on V3–10 is not on the neural arch but on the centrum (V9–10) or parapophysis (V3–8). The proximal ends of some anterior epineurals also are displaced ventrally in a few nonacanthomorphs. Among osteoglossomorphs, most epineurals originate on the centrum in mormyrids and in the notopterids *Xenomystus* (Table 3) and *Papyrocranus.* Among the elopomorphs in our sample, a few anterior epineurals originate on the centrum in *Notacanthus,* and among clupeomorphs about 15 do so in *Anchoa* (notation B_2 in Table 3). In euteleosts, *Esox* has 20 or more epineurals originating on the centrum, from about V10 onward (Table 4), and the aulopiforms *Scopelosaurus, Ahliesaurus* (both Notosudidae), *Scopelarchoides* (Scopelarchidae), *Coccorella* (Evermanellidae), *Bathysaurus,* and *Gigantura* have a number of epineurals originating on the centrum or (in *Scopelosaurus*) on the parapophysis (Table 5).

EPIPLEURALS.—As noted above, the epipleural series generally is easily distinguished from epicentrals and epineurals by ontogeny and distribution; epineurals and epicentrals develop in a rostrocaudal gradient, first appearing and being most strongly developed on anterior vertebrae, whereas epipleurals develop both rostrally and caudally from about the level of the first caudal vertebra.

Epipleural bones are absent in the most primitive teleosts, such as Mesozoic pholidophorids, early Jurassic leptolepids, and the Mesozoic ichthyodectiforms. When epipleurals first appear in the geological record, in late Jurassic "leptolepids," such as *Ascalabos, Tharsis,* and *Todiltia,* they are confined to the last few abdominal and first few caudal vertebrae. In small *Todiltia* (37 mm SL), for example, there are five pairs of epipleurals in that region, and in the largest specimens (60–80 mm SL) there are seven or eight (Schaeffer and Patterson, 1984:54). *Ascalabos* and *Tharsis* have about six epipleurals in that region, *Anaethalion* has about 12, and *Elops* has up to about 40 (Taverne, 1974). Osteoglossomorphs generally lack epipleurals, but in *Heterotis* they occur on up to 10 vertebrae (V26–36, Taverne, 1977b; cf. Table 3), and notopterids have been described as having a long series of them (Taverne, 1978). However, we interpret notopterid "epipleurals" differently; we consider them to be myorhabdoi (see section on osteoglossomorphs, below). Epipleurals generally are present in elopomorphs (elopids, megalopids, albulids, halosaurs, notacanths, anguilliforms) and in clupeomorphs (*Denticeps*, clupeoids) (Table 3); in clupeoids, the epipleural series may extend along the entire vertebral column (Table 3, *Chirocentrus;* Figure 7A). In euteleosteans, epipleurals occur as bones in gonorynchiforms, cypriniforms, characiforms, esocoids, alepocephaloids, stomiiforms, aulopiforms (where they always extend forward to V2 and often to V1), myctophiforms, and *Polymixia* (Tables 4, 5, 7). In many of these groups some epipleurals are forked proximally, like the epineurals above them (e.g., Plate 1E,G). When epipleurals have a proximal fork, orientation of the branches is the reverse of that in epineurals. There is an anteromedial branch, representing the primitive connection with the axial skeleton (rib, parapophysis, or haemal spine), and an anterodorsal branch that passes forward in the myoseptum, roughly parallel to the vertebral column, into the hypaxial forward flexure of the myomere. Toward the rostral and caudal ends of the epipleural series, the anteromedial branch is often absent, so that anterior and posterior epipleurals are rod-like bones, unconnected with the axial skeleton, representing the anterodorsal branch and the posterior body of the bone (notation D_0 in Tables 2–5).

Epipleural bones are absent in *Hiodon* and *Osteoglossum* (Table 3), *Dallia* and *Novumbra* (Table 4), and salmonoids (Sanford, 1990; Table 4) and osmeroids (McDowall, 1969, 1971, 1976; Rosen, 1974; Table 4; the osmeroid *Spirinchus* and the galaxiines *Galaxias* and *Paragalaxias* are exceptions, McDowall, 1969, 1978). In *Hiodon, Osteoglossum, Dallia, Novumbra,* and in all salmonoids and osmeroids (except *Lepidogalaxias, Stokellia,* and *Retropinna*) that we examined, we found the epipleural series to be represented by a series of ligaments (Tables 3, 4), with broadly the same distribution and attachment to the axial skeleton as the epipleural series of bones in other lower teleosts. There is a similar series of ligaments in the stomiiform *Pollichthys* (Table 5). Furthermore, in lower teleosts with epipleural bones, the series may be extended rostrally (e.g., *Megalops, Albula, Notacanthus, Anguilla, Esox,* Tables 3, 4, Plate 1A) or caudally (e.g., *Diplophos* and many aulopiforms, Table 5, Plate 2G) by a series of ligaments without ossification. To our knowledge, these epipleural ligaments have not been recorded previously.

Presence of epipleural bones evidently is a relatively derived feature in teleosts, because these bones are absent in pholidophorids, *Proleptolepis, Leptolepis* sensu stricto, and ichthyodectiforms (Patterson and Rosen, 1977). Among Recent osteoglossomorphs, they are absent except in *Heterotis* (we interpret the "epipleurals" of notopterids as myorhabdoi; see below in section on osteoglossomorphs). The epipleurals of *Heterotis* have the same form as in other teleosts; they are rod-like bones in direct continuity with the epipleural ligament (as in Plate 2G,H). Epipleural bones also are absent in Mesozoic osteoglossomorphs, such as *Lycoptera, Paralycoptera, Plesiolycoptera, Asiatolepis, Jiaohichthys, Yanbiania,* and the Huashiidae (*Huashia, Kuntulunia*) (Chang and Chou, 1976, 1977; Ma, 1983, 1986; Li, 1987). According to Chang and Chou (1977), some of these Mesozoic forms (*Asiatolepis, Paralycoptera*) are more closely related to osteoglossoids than to hiodontids, whereas some (*Lycoptera, Plesiolycoptera*) are immediately related to hiodontids. According to Ma (1983, 1986), other Mesozoic osteoglossomorphs (huashiids, *Jiaohichthys*) are incertae sedis within the group. A parsimony argument therefore implies that absence of epipleural bones is primitive for osteoglossomorphs, but that is inconsistent with Patterson and Rosen's (1977) and Schaeffer and Patterson's (1984) use of the presence of a few epipleurals in the middle of the trunk as a character grouping *Tharsis* (and possibly other Mesozoic stem-group teleosts, e.g., *Ascalabos* and *Todiltia*) with all Recent teleosts (osteoglossomorphs, elopocephalans). Because *Hiodon* and osteoglossoids have a series of epipleural ligaments, we assume that they also were present in Mesozoic stem-group teleosts and were the site of ossification of epipleural bones when they first appeared. We therefore regard the presence of a series of epipleural ligaments as primitive for teleosts (we have not found them in *Amia* or *Lepisosteus*), but because they are not preserved in fossils we cannot say when they first occurred. There are two possible interpretations of the distribution of ossification within those ligaments. Either it is synapomorphous for all taxa possessing epipleural bones, or it is homoplastic and has occurred independently in more than one lineage. The first interpretation (as in Patterson and Rosen, 1977) requires multiple losses in osteoglossomorphs (in *Arapaima,* the osteoglossid lineage, and in notopteroids according to Lauder and Liem's (1983, fig. 19) cladogram). The second interpretation requires multiple gains (*Heterotis,* elopocephalans, and at least one in "leptolepids"). We are unable to choose between the alternatives.

Patterson and Rosen (1977:126) and Lauder and Liem (1983) used presence of a long series of epipleural bones as a character of Elopocephala. The character remains uncontradicted (Tables 3, 4), and we add to it here the presence of proximally forked (with anterodorsal branches) epipleurals along some part of the series. Within elopocephalans, it is clear that epipleural bones have regressed, leaving only ligaments, on more than one occasion (e.g., the esocoid *Novumbra* and the stomiiform *Pollichthys,* Tables 4, 5). A decision on whether absence of epipleural bones in salmonoids and osmeroids (Table 4) is primary (primitive) or due to secondary loss can only be based on parsimony. Because they occur in elopomorphs, clupeomorphs, esocoids, ostariophysans, stomiiforms, and aulopiforms, we infer that epipleural bones were primitively present in elopocephalans, clupeocephalans, euteleosts, and neoteleosts, and that their absence in salmonoids and osmeroids is therefore secondary.

A more-derived state is complete absence of the epipleural series (no ligaments or bones). In nonacanthomorphs that we checked, epipleurals are completely absent only in *Lepidogalaxias, Stokellia,* and *Retropinna.*

The epipleural series of aulopiforms (Table 5) shows remarkable modifications that are discussed in detail in the section on aulopiforms below (p. 28). In brief, all aulopiforms are uniquely characterized by having epipleurals that extend forward to V1 or V2, and available evidence indicates that they have a different developmental pattern from that in other teleosts, where epipleurals develop rostrally and caudally from the region of the first caudal vertebra. In aulopiforms, epipleurals develop caudally from V3, V2, or V1. Furthermore, in most aulopiforms (all except *Bathypterois* and *Parasudis* in our sample) several or many of the anterior epipleurals have moved dorsally into the horizontal septum (notation $\mathbf{B}_3$ in Table 5). As a consequence, the epicentral series is truncated rostrally in various aulopiforms.

Alepocephaloids have epipleurals that extend unusually far forward, to V3, and in at least some alepocephaloids the distribution of epipleurals (virtually confined to the abdominal region) is such that they evidently do not develop from around the first caudal vertebra but from a more anterior focus. Data on development are lacking for alepocephaloids.

EPICENTRALS.—As bones, epicentrals are the most restricted of the three classes of intermusculars in lower teleosts. They lie in the transverse septum, and, like epineurals, they generally develop in a rostrocaudal gradient, from the anterior vertebrae backward. Unlike epineurals and epipleurals, they are almost always simple rods, without the proximal or distal forking that occurs in the other two series (we have observed distal forking of epicentral bones only in *Chanos* and kneriid gonorynchiforms). Epicentral bones do not occur in osteoglossomorphs, except in the notopterid *Notopterus* where the epicentrals are Taverne's (1978) "supraventrales"; they extend from the occiput to about V30 (see the section on osteoglossomorphs, below). Among elopomorphs, epicentral bones occur only in *Megalops* (Table 3; Forey, 1973; Rosen, 1985, fig. 9A). They occur in all clupeomorphs (Table 3; Greenwood, 1968 on *Denticeps;* Owen, 1866, Phillips, 1942, Yabumoto and Uyeno, 1982, and Whitehead and Teugels, 1985 on clupeoids; Greenwood, and Whitehead and Teugels call the epicentrals epipleurals). In many clupeoids there is a broad, vertical chevron of cartilage at the distal end of the anterior epicentral bones (Figure 7A′, Plate 1G, and below). Among ostariophysans, epicentrals occur as bones in all gonorynchiforms except the diminutive *Grasseichthys* (Table 4; Patterson, Fink, and Fink, ms) and in gymnotoids (Mago Leccia, 1978, as "costillas dorsales"; our observations on cleared-and-stained specimens of *Eigenmannia, Gymnotus, Hypopygus,* and *Sternopygus*). Among aulopiforms, ossified epicentrals occur in the chlorophthalmid *Parasudis,* in *Alepisaurus* (Table 5), and in *Omosudis.*

As with epipleurals, when epicentral bones are absent in nonacanthomorphs, we find a series of ligaments with broadly the same distribution as the bones. These are the anterior oblique tendons (AOTs) of Kafuku (1950), who reported their presence in *Anguilla,* clupeoids, ostariophysans, *Oncorhynchus, Plecoglossus,* and *Saurida* among nonacanthomorphs. Kafuku did not discuss the distribution of epicentral ligaments (AOTs) along the vertebral column of those fishes. As Tables 3-5 show, epicentral ligaments often begin on the occiput, attach to the centrum, parapophysis, or rib head on abdominal vertebrae, and to the centrum on caudal vertebrae. The only fishes in Tables 3–5 in which we failed to find epicentral bones or ligaments are the umbrid *Novumbra* (of which we had only poor specimens) and derived aulopiforms, such as paralepidids, *Bathysaurus,* and *Gigantura* (Table 5 and below).

The primitive pattern of epicentral ligaments and/or bones is a rostrocaudal gradient, with the structures extending caudally from the occiput or anterior vertebrae. Major exceptions to that pattern are found in *Umbra* and many aulopiforms. In *Umbra krameri* (Table 4) the epicentral series is interrupted, with ligaments from the occiput to V5 (the last one or two weaker than their predecessors) and then a gap back to V17; in *U. pygmaea* we failed to find epicentral ligaments anteriorly and the series appears to begin at V18. In many aulopiforms (Table 5) the epicentral series is truncated rostrally.

Anterior epicentral ligaments contain distal cartilaginous rods, similar to those of *Polymixia,* in many salmonoids and most osmeroids (Table 4). In salmonoids, the cartilage rods occur both in coregonids (*Coregonus lavaretus, Prosopium williamsoni, Stenodus leucichthys*) and salmonids (*Salvelinus alpinus, Salvelinus malma, Salmo salar, Salmo trutta,* Figure 7B, Plate 1D, *Thymallus thymallus,* see below), implying that they are primitive for the group according to Sanford's (1990) cladogram. Their ontogeny is well known in *Salmo* (Emelianov, 1935). In the salmonid *Thymallus,* we have observed cartilaginous rods in about the first 10 epicentral ligaments in small specimens of *T. thymallus* (40–45 mm SL), but they are not evident in larger specimens (~70–100 mm SL) of that species and of *T. arcticus.* In large *Thymallus arcticus,* Norden (1961) described and illustrated a series of "epipleural" bones that originate at the bases of the first eight ribs (V3–10) and are continuous with the head of the rib (like the anterior epicentrals in *Clupea,* Figure 7A). In our small specimens of *Thymallus* these bones are absent, and in larger specimens they are variable in number, and some or all of them may be independent of the rib (Table 4). We interpret the bones as epicentrals and as autapomorphous for *Thymallus* within salmonoids. (Norden (1961:736) reported a similar series of bones, apparently fused to the rib, in five out of 14 specimens of *Salvelinus fontinalis,* where they occurred on from 15 to 23 ribs. They were absent in other species of *Salvelinus.* The *Salvelinus* we examined (*S. alpinus* and *S. malma*) have cartilaginous rods in the epicentral ligaments and no epicentral bones.) In osmeroids, cartilaginous epicentrals occur both in northern taxa (Osmeroidea of Begle, 1991; *Hypomesus, Mallotus, Plecoglossus,* Figure 7C, in which they are particularly well developed) and in southern taxa (Galaxioidea of Begle, 1991; *Galaxias,* Plate 1H, *Prototroctes, Retropinna, Stokellia*). As with salmonoids, the distribution of cartilaginous epicentrals in osmeroids implies that they are primitive for the group according to Begle's (1991) cladogram. In *Lepidogalaxias,* which Begle (1991) placed as the sister group of Salangidae + Galaxiidae, the first six epicentral ligaments contain bone distally (Table 4), and the bone has the form of the epicentral

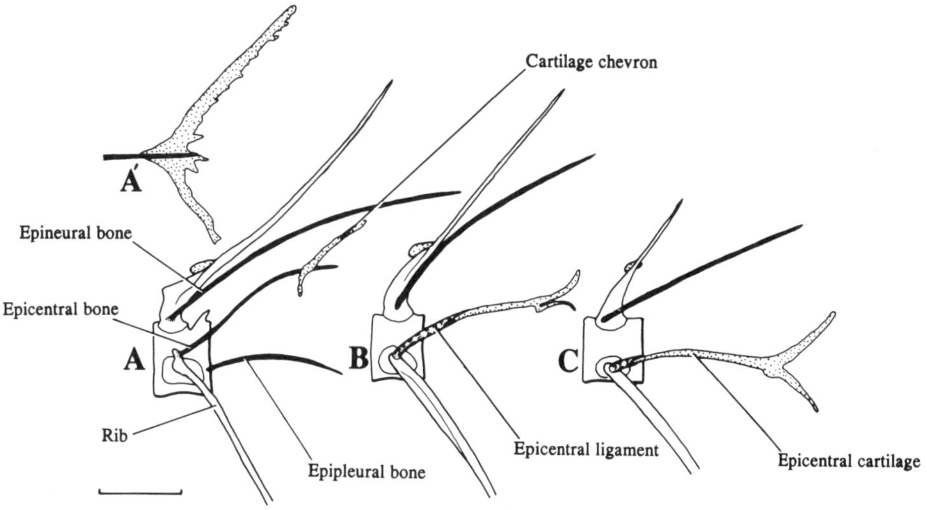

FIGURE 7.—Anterior vertebra and its appendages with associated epicentral cartilages in A and A′, *Clupea harengus* L.; B, *Salmo salar* L.; and C, *Plecoglossus altivelis* Temminck and Schlegel. Cartilage is stippled and intermuscular bones are in solid black. A, Fourth vertebra of BMNH 1970.2.17.22, ~ 65 mm SL. In *Clupea*, epicentrals are fused with the head of the pleural rib on anterior vertebrae and do not become independent until about the 20th vertebra. Anterior epipleurals lie free in the musculature (notation B_0, as in Table 3) so that the apparent epicentral position of the epipleural is illusory. A′, Tip of epicentral and associated cartilage chevron, after Göppert (1895, fig. 4; Göppert gave no scale or indication of size of fish from which he made the preparation). B, Fifth vertebra of BMNH 1981.9.22.11, 57 mm SL. C, Fourth vertebra of BMNH 1984.12.6.16, ~ 65 mm SL. (Scale bar 1 mm.)

cartilages in other osmeroids and in salmonoids, i.e., it is superficial, irregular in outline, and forked distally. We predict that it ossifies as cartilage bone, and it is an autapomorphic state of the osmeroid/salmonoid cartilaginous epicentral. Cartilaginous epicentrals also occur in the stomiiform *Maurolicus,* but not in the other stomiiforms that we have checked (Table 5).

Ontogeny of the epicentral cartilages in *Salmo* was the basis for Emelianov's (1935) distinction of ventral (our pleural) ribs from dorsal ribs (our epicentrals) by the criterion of centrifugal development (ventral ribs) versus centripetal development (dorsal ribs). Epicentral cartilages of *Salmo* first appear at about 30 mm TL as cartilaginous anlagen in the distal part of the horizontal septum, immediately beneath the lateral line nerve (Emelianov, 1935, fig. 20). They later extend medially toward the vertebrae as cartilage rods in the horizontal septum (Figure 7B); *Lepidogalaxias* (above) is the only fish known to us in which epicentral cartilages ossify. Emelianov regarded the ossifed epicentral of clupeoids (*Clupea, Alosa*) and the associated distal cartilage chevron (Figure 7A,A′) as together homologous with the cartilaginous epicentral in *Salmo.* But our observations of the clupeoid cartilage chevron indicate that there is no histological continuity or structural association between it and the epicentral bone, whose distal tip is merely surrounded by the cartilage, not in continuity with it. In salmonoids and osmeroids, the distal part of the cartilaginous epicentral is bifid, resembling the superficial cartilage chevron in clupeoids (Figure 7A′,B,C, Plate 1D,G,H).

In summary, there are four kinds of epicentrals: (1) epicentral ligaments, which we take to be the primitive teleostean condition; (2) epicentral bones (*Notopterus, Megalops,* clupeomorphs, gonorynchiforms, *Thymallus, Parasudis, Alepisaurus, Omosudis,* gymnotoids), which are membrane bones within the epicentral ligament; (3) ligaments containing rod-like epicentral cartilages (salmonoids, osmeroids, *Maurolicus, Polymixia*); (4) superficial ossifications in the distal part of the ligament (*Lepidogalaxias*), which we take to be ossified epicentral cartilages. Epicentral cartilages (3) develop centripetally, whereas epicentral bones (2), judging by their configuration, develop centrifugally except in the special case of *Lepidogalaxias* (4). The cartilaginous chevrons associated with the tips of epicentral bones in clupeoids are merely superficial and are not associated with epicentral ligaments. Because epicentral bones and cartilage chevrons coexist in clupeoids, with no structural or developmental continuity between them, they cannot be homologous. Emelianov (1935) therefore was mistaken in using the clupeoid condition as an intermediate stage in generalising the centripetal development of the epicentral cartilages in *Salmo;* the antithetic states of centripetally developed cartilage (e.g., salmonoids, *Polymixia*) and centrifugally developed bone (e.g., gonorynchiforms, *Parasudis*) imply that these two kinds of epicentrals (2, 3) are not homologous.

POSTERIOR OBLIQUE TENDONS (POTs of Kafuku, 1950).—Among lower teleosts considered in this section, Kafuku

(1950) recorded POTs in one elopomorph (*Anguilla*), three clupeomorphs (*Clupea, Sardinella, Clupanodon*), several ostariophysans (*Chanos*, three cyprinids, a cobitid), a salmonoid (*Oncorhynchus*), an osmeroid (*Plecoglossus*), and an aulopiform (*Saurida*). We confirm Kafuku's observation that in *Cyprinus* POTs penetrate the intervening epicentral ligaments en route to their insertion in the lateralis muscles.

Our most complete observations of POTs in lower teleosts were in the 164 mm SL *Albula* recorded in Table 3; significantly, this is about the largest of our cleared-and-stained specimens. In this fish the POTs in the abdominal region originate below the epicentral ligaments, from the ribs anteriorly and the parapophyses posteriorly. Each of them fans out immediately into a large number of small ligaments, so that in the abdominal region the horizontal septum is filled by evenly spaced, close-packed subdivisions of the POTs, all running obliquely forward at an acute angle to the long axis of the fish. Each epicentral ligament is crossed by the subdivided POTs originating on about 10 vertebrae behind that on which it originates. The epicentral ligaments lie on the upper side of the horizontal septum, dorsal to the POTs, which lie in that septum, and the segmental vessels run out on the underside of the septum, beneath the POTs. In the caudal region, POTs become less subdivided, more coherent and so easier to individualize. As an example, the epicentral ligament originating on V51 (PU21) is attached distally to subdivisions of the POTs from V58 and V59 (PU14–13). The last epicentral ligament originates on V53 (PU19). It is short and is attached distally to the innermost branch of the POT from V59. We have observed POTs in *Hiodon, Osteoglossum*, notopterids, *Elops, Anguilla*, clupeoids, *Umbra, Chanos*, cypriniforms, and stomiiforms, but not in *Esox* or *Gonorynchus* and in no salmonoids or osmeroids, no aulopiforms (except *Synodus*), and no myctophiforms. We failed to find POTs in small cleared-and-stained *Amia* and *Lepisosteus*. However, Westneat et al. (1993) reported POTs (and AOTs or epicentral ligaments) in the salmonid *Oncorhynchus*, in *Amia, Lepisosteus*, and even in sharks (*Carcharhinus*). They gave no size range for the fishes examinined.

MYORHABDOI.—In various lower teleosts there are one or two extra series of unattached intermuscular bones, developed respectively in the uppermost and lowermost forward flexures of the myosepta. These bones, which were named myorhabdoi by Chapman (1944), occur, for example, in *Notopterus* (Taverne, 1978, and the section on osteoglossomorphs, below), some species of the congrid eel *Ariosoma* (Smith 1989, figs. 516, 517), many clupeoids (which generally have only the dorsal series, or have that series much better developed than the ventral series; Grande, 1985, figs. 10, 11; Chapman, 1944, fig. 14), gymnotoids (Mago Leccia, 1978; our observations on several genera), and the stomiiform *Triplophos*. Notopterids have two additional series that we interpret as myorhabdoi: one in the epineural position, following behind the true or attached epineurals, and one in the epipleural position (see the section on osteoglossomorphs, below). *Notopterus* has two further series of myorhabdoi: one (Taverne's, 1978, "centro-vertébrales") oriented obliquely posteroventrally between the epineurals and epicentrals on about V1–25, and one associated with the anterior epicentrals, on the occiput and about V1–5 (see the section on osteoglossomorphs, below). In our opinion, myorhabdoi are adventitious structures, typical of long-bodied, laterally compressed fishes. *Anchoa* and *Xenomystus* are the only tabulated genera that have them, and we have entered them (notation M, Table 3) only in the latter.

OCCIPUT, CRANIAL RIBS AND INTERMUSCULARS, AND ACCESSORY NEURAL ARCH.—There has been much debate, historical and recent, over the composition of the occiput in teleosts. It frequently has been asserted that one or more vertebrae, or a "half-centrum," have fused with the occiput in one teleost or another. Rosen (1985) gives the most recent review of the question.

Allis (1898, 1903) showed that in *Scomber* the configuration of the anteriormost myotomes and the attachment of their myosepta to the occiput are the same as in *Amia*, given that in *Amia* two vertebrae fuse with the occiput during ontogeny. One useful landmark in homologising occipital structures is the internal limb of the posttemporal, which lies in the first myoseptum (between myomeres 1 and 2) and attaches to the intercalar proximally. Among teleosts surveyed in Tables 3–5, the only one with an intermuscular originating on the intercalar is *Gonorynchus* (En Ic, Figure 4). There, the bone is in series with the epineurals, and, together with the internal limb of the posttemporal (represented only in ligament in *Gonorynchus*), it passes out toward the posttemporal and must lie in the myoseptum. In *Scomber*, the second and third myosepta attach to the occiput close behind (i.e., posteromedial to) the first myoseptum, and Baudelot's ligament lies in the third myoseptum; the fourth myoseptum attaches to the first vertebra and contains the first epineural (Allis, 1898, 1903). The match between anterior myosepta in *Amia* and *Scomber* (Allis, 1898) implies that the pattern in *Scomber* is general for teleosts. If so, no teleost should have more than three serially homologous intermusculars on the braincase, unless one or more vertebrae are fused into the occiput, and none we know of has more than three. Among lower teleosts surveyed in Tables 3–5, *Albula, Chanos*, and *Gonorynchus* have three such bones. In *Albula*, the three are in series with the epineurals; the first two are branched proximally and extend into the posttemporal fossa without direct attachment to the braincase, and the third attaches to the exoccipital by ligament. They must lie in myosepta 1–3. In *Chanos*, the three bones are also in series with the epineurals, but they represent only the anteroventral branches of those bones; like the first two bones in *Albula*, they extend into the posttemporal fossa without direct attachment to the braincase, and they must lie in myosepta 1–3. In *Gonorynchus*, the first of the three bones is an epineural in myoseptum 1 (above), and the other two each represent fusion between an epicentral and the anteroventral branch of an epineural; these two originate respectively on the exoccipital and basioccipital and must lie in myosepta 2 and 3. There is thus no evidence from the segmental intermusculars that any teleost has a vertebra fused into the braincase, and the constancy of origin of the first rib on V3 (Tables 3–5) is corroborative. In fact, among lower teleosts that we survey in Tables 3–5, there is evidence of ontogenetic fusion of a vertebra into the occiput only in *Heterotis* (Daget and D'Aubenton, 1957). That fusion is reflected in the ligaments and ribs, because both Baudelot's ligaments (double in *Heterotis*) originate on the braincase, and the first rib is on V2. In one specimen of *Heterotis*, the BMNH skull described by Ridewood (1905), the fusion is incomplete, and the second Baudelot's ligament (ossified) remains on the vertebral centrum. The example of *Heterotis* implies that where a vertebra fuses with the braincase in lower teleosts, it will carry Baudelot's ligament. In *Esox*, Jollie (1975:78) wrote that the basioccipital "includes the first vertebral body," but he presented no developmental evidence for this apart from presence of an accessory neural arch (see below). Baudelot's ligament is on V1 in *Esox* (Figure 8, Table 4).

Fink and Fink (1981:326) pointed out that an unattached neural arch in front of that of V1 (Rosen's, 1985, accessory neural arch; abbreviated ANA here and in Tables 3–5) probably is primitive for teleosts and is absent in all ostariophysans. They recorded presence of ANA in *Elops*, the clupeoids *Dorosoma, Harengula,* and *Anchoa,* and in *Esox* and the stomiiform *Diplophos.* Tables 3 and 4 show that ANA is absent in the osteoglossomorphs in our sample, absent in *Megalops, Albula, Halosaurus, Notacanthus,* and *Anguilla* amongst elopomorphs, present in *Clupea* and *Chirocentrus,* but absent in *Denticeps, Clupeichthys,* and *Jenkinsia* amongst the clupeomorphs in our sample, absent in *Umbra* and *Novumbra,* and, according to our sample, is present in salmonids, alepocephaloids, and northern osmeroids, but absent in argentinoids and southern osmeroids. In stomiiforms, ANA is present in a number of genera (in addition to *Diplophos*), which are rated as primitive members of the group on other criteria (A.S. Harold, pers. comm., 1993). Among aulopiforms (Table 5), ANA is present in *Aulopus,* synodontids, and harpadontids but is absent elsewhere. It does not occur in ctenosquamates (myctophiforms plus acanthomorphs). We note that Rosen (1985, fig. 9) illustrated an ANA in *Albula,* but we have examined several specimens, including an ontogenetic series, and have not found the structure.

Fink and Fink (1981:326) also reported ANA in *Polypterus* and *Amia,* implying that the structure is primitive for actinopterygians. There is no ANA in those fishes, because the occipital neural arch(es) is associated with vertebrae fused into the occiput during ontogeny (Jollie, 1984a, 1984b). We have not observed ANA, nor is one reported, in Mesozoic stem group teleosts such as pholidophorids, leptolepids, and ichthyodectiforms.

Given the mosaic pattern of presence and absence of ANA in lower teleostean groups, there are two possible interpretations: either it is synapomorphous at some level and has been lost repeatedly, or it has arisen repeatedly and is nonhomologous from group to group. If it is synapomorphous, according to the information summarized above and in our tables, it appears to be an elopocephalan character, as it is not recorded in osteoglossomorphs or nonteleostean actinopterygians. Treating ANA as an elopocephalan character would require that it be lost at least 10 times (at least once in elopomorphs, at least twice in clupeomorphs, and in at least seven different euteleostean lineages—ostariophysans, esocoids, argentinoids, osmeroids, stomiiforms, aulopiforms, ctenosquamates). Treating ANA as independently acquired demands that it arose once in *Elops,* at least once in clupeoids, in *Esox,* in alepocephaloids, in salmonoids and/or osmeroids, and in stomiiforms and/or aulopiforms, with subsequent loss within several of those groups. A choice between these alternatives ultimately will have to be made on grounds of parsimony, presumably with differential weighting for gain and loss of ANA. We need a more robust cladogram of teleosts and a more extensive survey for ANA before any rational decision is possible, but in our analysis of aulopiforms (Figure 11, Table 6) ANA was resolved as a derived character.

We have no useful conjecture on the origin of ANA, although we are convinced, for reasons discussed at the beginning of this section, that it does not represent the remains of a missing vertebra.

BAUDELOT'S LIGAMENT.—Baudelot's ligament is the name now given to the stout ligament that originates on the ventrolateral aspect of an anterior vertebra (usually the first) or the occiput (usually the basioccipital) proximally and inserts on the supracleithrum distally, usually after looping around the upper part of the cleithrum. As an aside, we think it worth clarifying the origin and history of the term, so far as we have been able to understand it.

Baudelot (1868:83–84) seems to have been the first to notice the ligament, and he named it the "*scapulo-vertébral*" ligament. That name was used by subsequent late nineteenth- and early twentieth-century European authors (e.g., Siebenrock, 1901), and it was used in the early 1960s by Monod (1960:631, with a reference to Baudelot, 1868; 1963:271, as "*ligamentum scapulo-vertébrale* de BAUDELOT"). Baudelot's term was not used by early twentieth-century Anglophone authors, such as Starks (1904b:258–259; 1923:267) or Allis (1903:103, 110; 1909:36, 57, as "occipito-supraclavicular ligament"). In 1962, Weitzman (1962:42) called it the "trans-scapular ligament," and Rosen (1962:8) called it the "ligamentous support." Thus, the name "Baudelot's ligament" does not seem to have been current in the early 1960s. The earliest use of the term that we have found is in Rosen (1964:241) when he stated that "the shoulder girdle is attached by a ligament (Baudelot's ligament) to the basicranium." In Greenwood et al. (1966) the term "Baudelot's ligament" is used several times (pp. 352, 353, 354, 388, 389, 392), but on p. 386,

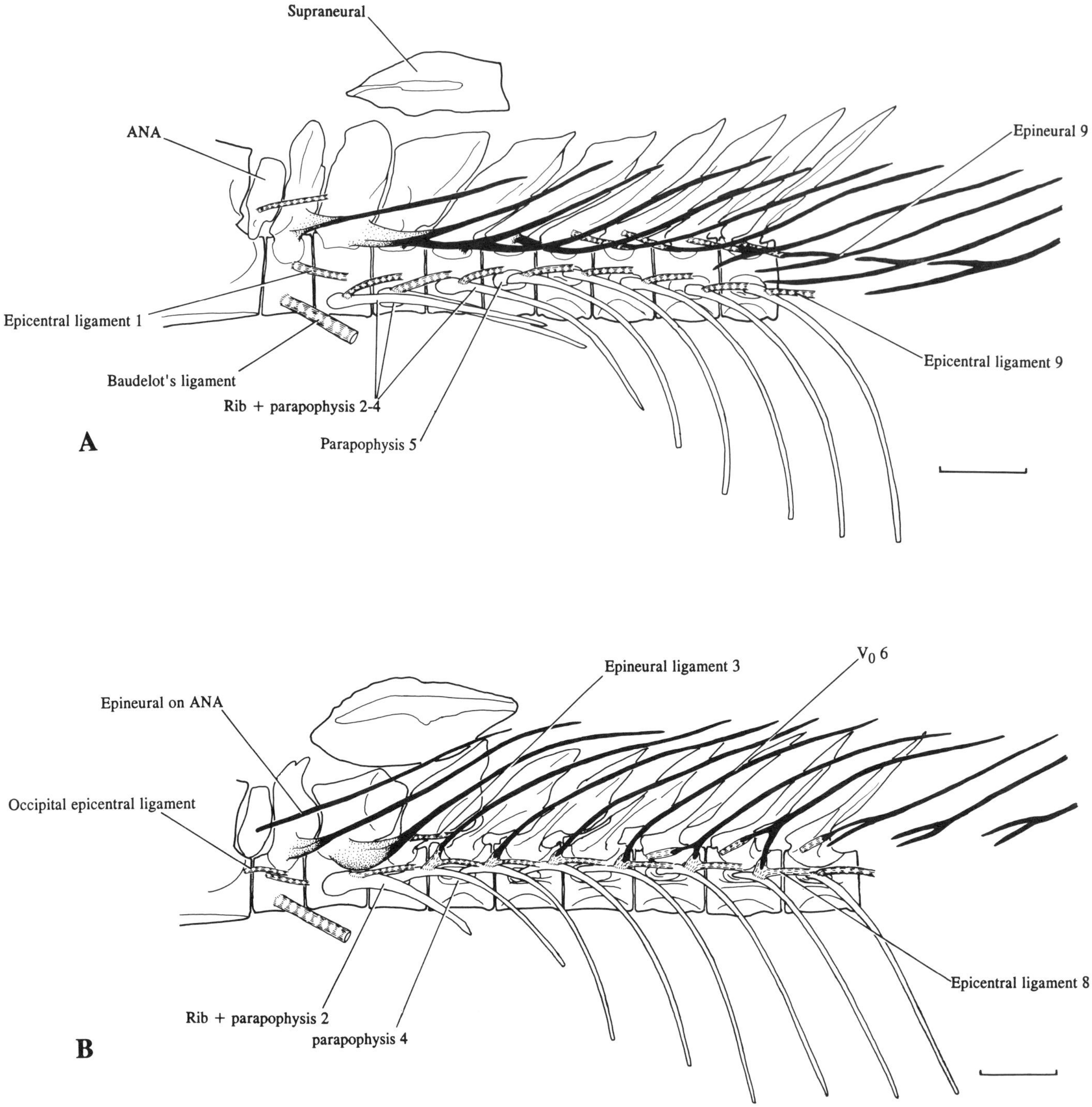

FIGURE 8.—Anterior vertebrae and associated structures in A, *Esox lucius* L., ROM 598 CS, 62 mm SL; B, *Esox americanus* Gmelin, BMNH 1982.11.60.16, 61 mm SL. The intermuscular bones are in solid black, except for proximal parts of anterior epineurals that are broadly fused to neural arches, where fusion is indicated by stipple. The ligaments are shaded. (Scale bars 1 mm.) (Abbreviations: ANA, accessory neural arch; V_0, intermuscular representing unattached anteroventral limb and posterior body of an epineural. Intermusculars and ribs are numbered or labeled according to vertebra (number) with which they are associated.)

in a part written by Weitzman, the structure is called the "transscapular ligament," as in Weitzman's earlier (1962) work. Our guess is that the term "Baudelot's ligament" was coined by Rosen (1964:241), perhaps intentionally but more probably through misreading of Monod's (1963:271) phrase "*ligamentum scapulo-vertébrale* de BAUDELOT." The term soon gained general acceptance through repeated use in the widely read paper by Greenwood et al. (1966).

Tables 3-5 show that the origin of Baudelot's ligament on V1 is remarkably constant in lower (nonacanthomorph) teleosts. The only exceptions that we have found are as follows. In the osteoglossomorphs *Heterotis* and *Osteoglossum* and the cobitid *Acanthopsoides* there are two ligaments, one on the basioccipital and one on V1 (that on V1 is ossified in *Heterotis*, in which V1 is fused with the basioccipital). In *Galaxias* there are also two ligaments, one from the basioccipital and one from V1, which join distally before attaching to the supracleithrum. In the characoid *Alestes*, the ligament is on the basioccipital; according to Fink and Fink (1981:335) this is true of all characiphysans (characiforms and siluriforms) and of several cypriniforms. In the paralepid aulopiforms *Paralepis* and *Macroparalepis* there is a ligament with an origin extending from V1 to V2, whereas the aulopiform *Alepisaurus* has the origin of the ligament extending from V2 to V4 (we have observed the same condition in *Omosudis*). The clupeoid *Chirocentrus* has an ossified ligament with a double attachment, to the anteroventral corner of V3 and to the posteroventral corner of V2. The ligament on V1 is ossified in the engraulid *Anchoa*, the stomiiform *Diplophos*, and the harpadontid aulopiforms *Harpadon* and *Saurida* among our sample. It also is ossified in *Dallia* (Nelson, 1972; Wilson and Veilleux, 1982). Finally, Baudelot's ligament is absent in *Notacanthus*, gonorynchiforms, the catostomid *Hypentilium*, and the aulopiform *Gigantura*; in *Anguilla* a sheet of ligament with origins on V1-5 attaches to the supracleithrum; some or all of these may be homologous with Baudelot's ligament.

RIBS.—Ribs of teleosts are primitively preformed in cartilage and ossify as cartilage bones, but in many teleosts some or all ribs ossify as membrane bone (Emelianov, 1935; Faruqi, 1935). Although our criterion for distinguishing ribs formed in cartilage from those formed in membrane bone (the configuration of the tip in cleared-and-stained specimens) is less refined in comparison with the serial sectioning used by Emelianov, Faruqi, and others, the survey in Tables 3-5 is probably the most extensive yet published. All ribs are preformed in cartilage in the osteoglossomorphs, elopomorphs, and clupeomorphs in our sample, with two exceptions. The first is a diminutive rib in membrane on V2 in our *Heterotis* and *Osteoglossum* and on V3 in *Xenomystus*. The second includes the elopomorphs *Halosaurus*, in which most ribs are in membrane but the last dozen are in ligament, *Notacanthus*, where all the ribs are in ligament, and *Anguilla*, where the first few ribs are in ligament and the remainder are in membrane. Among nonacanthomorph euteleosts, all the ribs are preformed in cartilage in *Chanos* and in the salmonoids, alepocephaloid, osmerids, retropinnid, galaxiid, stomiiforms, and myctophiforms in our sample (Tables 4, 5). In *Esox*, the first one (*E. americanus*) to three (*E. lucius*) ribs form in membrane bone and are coossified with the parapophysis, a configuration that occurs elsewhere (to our knowledge) only in the highly modified first two ribs (tripus, on V3, and transformator process, on V4) of otophysans. In *Umbra* and *Novumbra*, the first rib is in membrane and the rest (or all but the last) are in cartilage. Among gonorynchiforms, *Gonorynchus* has all but the first rib in membrane, whereas *Parakneria* has the first and about the last eight in membrane. The ribs (other than those on V3 and V4) are in cartilage in the otophysans in our sample, except in the cobitid *Acanthopsoides*, where they are all in bone. The argentinoid *Glossanodon* has about the last 10 ribs in membrane, and *Lepidogalaxias* has them all in membrane. In aulopiforms, the condition ranges from the last few ribs in membrane (*Parasudis, Scopelosaurus, Coccorella*) or only the first in membrane (*Bathysaurus*), through most of them in membrane (*Chlorophthalmus, Bathypterois, Ahliesaurus, Scopelarchoides*), to all ribs in membrane (the remainder of the genera sampled in Table 5).

Ribs of nonacanthomorph teleosts are, in our experience, restricted to the abdominal region, where they always lie in the peritoneal membrane.

As with the origin of Baudelot's ligament on V1, origin of the first rib on V3 is remarkably constant in lower teleosts. Exceptions in Tables 3-5 are the first rib on V1 in the aulopiforms *Paralepis and Alepisaurus* (and *Omosudis*); first rib on V2 in *Heterotis, Osteoglossum*, all esocoids, *Searsia, Osmerus, Lepidogalaxias, Galaxias, Bathypterois*, and *Harpadon;* on V4 in *Megalops* and *Pseudotrichonotus;* on V5 in *Halosaurus* and the aulopiforms *Macroparalepis, Synodus*, and *Trachinocephalus;* on V6 in *Elops;* and on V15 in *Notacanthus*, in which all ribs are in ligament. In *Paralepis, Alepisaurus*, and *Omosudis*, the only teleosts we have found with ribs on V1, the ribs on V1 and V2 are fused to the centrum (Table 5), another unique feature of these genera.

The "cranial ribs" of *Gonorynchus* (two pairs on the occiput, one on V1) lie in the peritoneal membrane, which is the position of ribs, but we do not believe that they are serially homologous with ribs. They appear to be an extra series of intermuscular bones, which is autapomorphous for gonorynchids.

COMMENTS ON INDIVIDUAL GROUPS AMONG LOWER TELEOSTS

In this section we comment on aspects of the intermuscular bones, ligaments, and cartilages in various major groups that may bear on relationships.

OSTEOGLOSSOMORPHS.—We recorded the intermuscular bones and ligaments in only four osteoglossomorphs, *Hiodon, Osteoglossum, Heterotis,* and *Xenomystus* (Table 3), and our specimens were far from ideal. Nevertheless, it is worth noting that *Hiodon* exhibits what we take to be the most primitive condition found in any Recent teleost, with all epineurals unbranched and fused to the neural arches proximally, epineural ligaments extending the epineural series caudally, a series of epicentral ligaments from the occiput back into the caudal region, and a short series of epipleural ligaments in the middle part of the trunk. *Osteoglossum, Heterotis,* and *Xenomystus* are more derived, with all (*Heterotis, Xenomystus*) or most (*Osteoglossum*) of the epineurals free from the neural arches, and *Heterotis* has all the epipleurals ossified whereas *Xenomystus* has two series of bones that we interpret as myorhabdoi (below), one succeeding the epineurals and one in the epipleural position. The osteoglossoid *Pantodon* agrees with *Hiodon* in having all the epineurals (except the first) fused with the neural arch (Taverne, 1978), but we have seen no good cleared-and-stained material of *Pantodon,* and so we cannot comment on its ligaments. *Arapaima* (Taverne, 1977b, and our observations) is derived in lacking epineurals on about the first 20 vertebrae, and in having the anterior epineurals on the parapophysis.

Among notopteroids, in addition to *Xenomystus* (Table 3), we examined specimens of mormyroids (*Gymnarchus, Isichthys, Marcusenius*) and the other two notopterid genera (*Notopterus, Papyrocranus*), and Taverne (1972, 1978) has given good descriptions of the osteology of all genera in the two groups. The two groups agree in having only a short series of epineural bones that are attached (not fused) to the neural arches (in *Notopterus* and *Gymnarchus*) or centra (in mormyrids, *Xenomystus,* and *Papyrocranus*): on V4 to about V15 in *Gymnarchus,* V1 to about V10 in mormyrids, V1 to about V13 in *Notopterus,* V1–16 in *Papyrocranus,* and V1–7 in *Xenomystus.* In mormyrids and notopterids (except *Notopterus*) there are epineural ligaments beyond the short series of attached epineural bones. In mormyrids there are no further epineural bones, but in *Xenomystus* and *Papyrocranus* each epineural ligament fans out distally, and beginning immediately behind the last attached epineural bone there is a long series of superficial, lath-like bones that are remote from the vertebrae and loosely attached to the diffuse ligaments; we interpret them as myorhabdoi (notation M in Table 3, and see below). In *Notopterus,* where there are no epineural ligaments behind the last attached epineural bone (~V13), there is an abrupt transition from attached epineural bones to superficial, unattached bones that are deeply subdivided or frayed anteriorly; as with the lath-like bones in other notopterids, we interpret them as myorhabdoi. A long series of epicentral ligaments and a short series of epipleural ligaments are present in mormyrids and notopterids, and in *Notopterus* the epicentrals are ossified on about V1–30 (Taverne's, 1978, "supraventrales," below).

In notopterids, Taverne (1978) described a long series of epipleural bones in each of the three genera. Our interpretation differs. The epipleurals of the osteoglossoid *Heterotis* have the same form as in other teleosts; they are rod-like bones, in direct continuity with the epipleural ligament (as in Plate 2H). But in notopterids they have a different configuration, and they differ between the three genera. In *Papyrocranus,* the most anterior "epipleural" is about at the level of the first caudal vertebra (~V16), and epipleural ligaments are present from about V14 to about V35. The "epipleural" bones are superficial and lath-like. The epipleural ligaments fan out distally, and, in segments with both "epipleurals" and ligaments, the diffuse ligament attaches loosely and broadly to the posterior end of the bone. In *Xenomystus* (Table 3), the "epipleural" bones extend forward to about V4, but, as in *Papyrocranus,* epipleural ligaments extend forward only to the antepenultimate abdominal vertebra (V9); they extend back to about V22. In segments with both epipleural ligaments and bones, the bones are shaped rather like a single-blade propellor, and, as in *Papyrocranus,* the epipleural ligaments fan out distally, but they attach broadly to the middle of the bone rather than to its posterior end. Thus, in *Xenomystus* and *Papyrocranus* the "epipleural bones" have the same relationships as the unattached "epineural bones" above them. Because the "epipleural bones" do not resemble the epipleurals in any other teleosts, and because we know of no other teleost in which there is an abrupt transition from an attached epineural on one vertebra to a superficial, lath-like epineural on the next, we believe that both series of bones are myorhabdoi (notation M in Table 3), which is autapomorphous for notopterids.

In *Notopterus,* the intermusculars are much more complex than in *Xenomystus* and *Papyrocranus.* Taverne (1978) described six series of epipleurals. The uppermost and lowermost series (Taverne's "supradorsales" and "infraventrales") resemble and correspond to the myorhabdoi of clupeoids and gymnotoids. As described above, behind the last attached epineural (~V13) there are no epineural ligaments, and there is an abrupt transition from attached epineural bones to unattached, superficial bones that we interpret as myorhabdoi, as in *Xenomystus* and *Papyrocranus.* The epicentrals are ossified from the occiput to about V30 (Taverne's "supraventrales"), and on the occiput and the first four or five vertebrae they are accompanied by extra ossifications (myorhabdoi), so that there may be up to four additional bones in the myoseptum lateral to the epicentral (Taverne, 1978, figs. 76, 77). Between the epicentrals and the epineurals, from the occiput to about V25, there is a further series of small, irregular, unattached intermusculars (Taverne's "centro-vertébrales") that are inclined posteroventrally. The anterior members of this series are more dorsally placed, rising upward medial to the epineurals and decreasing in size anteriorly. We take them to be another series of myorhabdoi, which is autapomorphous for *Notopterus.* Finally, *Notopterus* has the series of intermusculars described as epipleurals by Taverne (1978). They extend from about V8 back to the tail, and most of them resemble the epineural myorhabdoi above them, being brush-like or fimbriate at both ends (Taverne, 1978, figs. 78–80). There are epipleural ligaments from about V8 to V30. Most of these ligaments fan out distally and attach loosely or diffusely to the middle part of the "epipleurals," as in *Papyrocranus*, but the ligamentous attachment moves toward the anterior part of the bone in the anterior "epipleurals," and the first two or three members of the "epipleural" series (~V8–10) are small, rod-like bones attached to the ligament by their anterior tips. Thus, the first two or three members of this series resemble true epipleurals (as in Plate 2G,H or *Heterotis*, for example), but the great majority of them resemble the lath-like bones in the epipleural position in *Xenomystus* and *Papyrocranus* that we interpret as myorhabdoi. In other lower teleosts with epipleural bones, the ligamentous attachment is best developed in the middle of the series (examples in Tables 3, 4), not in its first few members. Thus, in *Notopterus*, as in *Xenomystus* and *Papyrocranus*, the "epipleurals" do not agree with those bones in other teleosts, and as in those genera, we interpret them as myorhabdoi. In short, among notopterids, *Xenomystus* and *Papyrocranus* have two series of myorhabdoi, one in the epineural and one in the epipleural position, whereas *Notopterus* has six series of them: one dorsal and one ventral, corresponding to the myorhabdoi of other teleosts (e.g., clupeoids, gymnotoids); one in the epineural and one in the epipleural position, corresponding to the myorhabdoi in other notopterids; and two autapomorphous series, one between the epicentrals and epineurals and one accompanying the epicentrals on the occiput and first few vertebrae.

Although there are reasons to believe that *Hiodon* is related to notopteroids, it generally is agreed that "of all osteoglossomorphs, the Hiodontidae show the greatest aggregate of primitive characters" (Greenwood, 1973:314). The intermusculars bear out this assessment.

ELOPOMORPHS.—We record the intermuscular bones and ligaments in *Elops, Megalops, Albula, Halosaurus, Notacanthus*, and *Anguilla* (Table 3). In comparison with the basic generality of *Hiodon*, these elopomorphs share several derived features with clupeomorphs and lower euteleosts: (1) some epineurals are not fused with the neural arch; (2) some epineurals are forked proximally (not found in *Anguilla*, but present in many other eels, e.g., Smith, 1989, figs. 516, 517); (3) there is an extensive series of epipleural bones, covering at least one-third of the abdominal region and most of the caudal region; (4) some epipleurals are forked proximally (again, not found in *Anguilla* but present in many other eels). Characters (2) and (4) almost always co-occur, but *Esox* (Table 4) shows that co-occurrence is not consistent.

In *Halosaurus*, the last dozen ribs are in ligament, whereas in *Notacanthus* all ribs are in ligament; this derived condition is probably synapomorphous for halosaurs and notacanths. *Albula, Notacanthus*, and *Anguilla* share an extensive series of anterior epipleural ligaments, extending forward to V10 in *Albula* and *Notacanthus*, and to V7 in *Anguilla*. In *Halosaurus* there are epipleural ligaments from V37 to V57 and bones thereafter.

CLUPEOMORPHS.—We record the intermusculars in *Denticeps* and five clupeoids, the clupeine *Clupea*, the pellonuline clupeid *Clupeichthys*, the dussumieriine clupeid *Jenkinsia*, the engraulid *Anchoa*, and the chirocentrid *Chirocentrus* (Table 3). All have ossified epicentrals, but in other respects *Denticeps* (the only member of the suborder Denticipitoidei) is very different from clupeoids. *Denticeps* lacks epineurals on V1–11, has epineural ligaments on V12–19, and the first ossified epineural on V20; these conditions are unique among the teleosts we have sampled, although *Arapaima*, with the first epineurals on the parapophyses of V21 (Taverne, 1977b:229) or V22–24 (pers. obs.) approaches it. The epipleural series in *Denticeps* is confined to the caudal region, whereas in clupeoids it extends very far forward (to V3 in *Clupeichthys*, to V2 in *Clupea, Jenkinsia*, and *Anchoa*, and even to the occiput in *Chirocentrus*, where there are two cranial epipleurals, each fused proximally with one of the two epicentrals on the occiput). The epicentral series in the three clupeids and *Chirocentrus* share one feature that is almost unique among the teleosts that we record in Tables 3–5: fusion of the epicentral with anterior ribs. That fusion occurs on about the first 20 ribs in *Clupea* and *Clupeichthys* (recorded in *Clupea* by Chabanaud, 1942; also present in *Sardinops*, Phillips, 1942; and observed by us in the clupeines *Harengula, Opisthonema*, and *Sprattus*), on about 8–10 ribs in *Jenkinsia* (as in the dussumieriine *Spratelloides;* observed on about 12–15 ribs in *Etrumeus*), and on 6–7 ribs in *Chirocentrus*. This rib/epicentral fusion also may occur on a few ribs in the salmonid *Thymallus* (Table 4 and below). It does not occur in pristigasterids that we checked (*Chirocentrodon, Ilisha, Odontognathus, Opisthopterus, Pellona*), or in other engraulids (*Engraulis, Cetengraulis*) or in the coiliids (sensu Grande and Nelson, 1985) *Setipinna* and *Thryssa;* the derived coiliid *Coilia* is an exception, for the first rib is enlarged and fused with the epicentral. The rib/epicentral fusion occurs on about the first 20 ribs in alosines (*Alosa, Brevoortia*) and on the first few ribs (up to about eight) in dorosomatines (*Anodontostoma, Dorosoma*). In *Clupea*, the fusion is ontogenetic, at about 40 mm TL (Ramanujam, 1929). It may not occur in the diminutive pellonuline *Sierrathrissa* (Whitehead and Teugels, 1985, figs. 33, 34), which reaches only 30 mm SL. This rib/epicentral fusion certainly is derived, and it suggests that chirocentrids and clupeids (clupeines, alosines, dorosomatines, dussumieriines, and pellonulines) form a monophyletic group. This would give further support to Grande's (1985, fig. 9) scheme, in which he cited only one character supporting that grouping (Clupeoidea), "increase in pleural rib to preural vertebrae ratio," a character that may not be derived.

Cartilage chevrons at the tips of epicentrals are recorded in all clupeoids except *Chirocentrus* in Table 3. They are found only in clupeoids among lower teleosts, although similar structures are associated with epineurals in holocentrids and the stephanoberyciform *Rondeletia* (Table 7 and below). Emelianov (1935) recorded cartilage chevrons in *Clupea* (Clupeinae) and *Alosa* (Alosinae); we have observed them in other clupeids (e.g., *Cynothrissa*, Plate 1G, *Pellonula*, *Sierrathrissa*) and engraulids (*Engraulis*, and in *Stolephorus* at 40 mm SL, although not in *Cetengraulis* at 55 mm) and in the coiliid *Thryssa* (although not in *Setipinna* at 80 mm SL). Lance Grande (pers. comm., 1992) has observed them in the clupeines *Harengula* and *Sardinella* and the dorosomatine *Dorosoma*, but not in his specimens of the clupeine *Lile*, the dorosomatine *Nematalosa*, the dussumieriine *Etrumeus*, or the coiliid *Coilia* (in our *Coilia*, BMNH 1979.8.24.133-4, ~110 mm SL, there are diminutive chevrons on the first five or six epicentrals in one specimen but none in the other). Cartilage chevrons do not occur in the pristigasterids we checked (*Chirocentrodon*, *Ilisha*, *Odontognathus*, *Opisthopterus*, *Pellona*) or in the three double-stained *Chirocentrus* that we have seen. They might be synapomorphous, indicating that Engrauloidea are more closely related to Clupeoidea than are Pristigasteroidea (cf. Grande, 1985, fig. 9), but their distribution, so far as we have been able to check, is not congruent with the epicentral/rib fusion discussed above, which groups *Chirocentrus* with clupeids. We have not seen enough double-stained clupeoids (the cartilage chevrons are detectable only with alcian blue) to resolve this conflict.

The intermusculars of clupeoids show some remarkable complications (e.g., Whitehead and Teugels, 1985, on *Sierrathrissa*) and a comprehensive study could well be profitable.

OSTARIOPHYSANS.—Fink and Fink (1981) gave further substance to the idea, first proposed by Rosen and Greenwood (1970), that gonorynchiforms are the sister group of otophysans; the two groups together make up the Ostariophysi. The anterior intermusculars, which were not considered by Fink and Fink (those of *Chanos* and *Gonorynchus* were discussed by Rosen and Greenwood, 1970), are relevant here. In Table 4 we record the intermusculars of four otophysans, a characiform (*Alestes*), and three cypriniforms: a cyprinid (*Hemibarbus*), a cobitid (*Acanthopsoides*), and a catostomid (*Hypentelium*). In these forms, intermusculars of the first four or five vertebrae are modified, evidently in association with development of the Weberian apparatus. In all four genera the epicentral series is interrupted or truncated anteriorly in the region of the Weberian apparatus. In all four genera the epineurals are forked proximally (back to about V30 in *Alestes*, *Hemibarbus*, and *Hypentelium*, and to about V25 in *Acanthopsoides*), and the anteromedial (primary) branch of the epineural is attached to the neural arch by ligament. There are no medial branches or ligaments anterior to V5 in *Alestes*, V6 in *Acanthopsoides*, and V12 in *Hypentelium*. But in *Hemibarbus*, epineurals are attached by ligament forward to V1 (except on V2, which lacks an epineural). Epineurals without medial branches or ligaments on anterior vertebrae are represented only by the anteroventral branch and the posterior body (notation V_0 in Table 4). The anterior tips of these epineurals tend to establish connections with other structures.

In *Alestes*, the epineural of V4 has an extra bifid or trifid ventral branch from its proximal end, with ligaments running down to the swimbladder from the tips of those branches, and the epineural of V3 has a ligament from its proximal end running to the anterior tip of the transverse process of the third neural arch (which is a structure unique to characiforms; character 71 of Fink and Fink, 1981:329). The anterior tips of the epineurals of V1 and V2 in *Alestes* are without obvious ligamentous attachment, and anterior to them there is one further epineural passing forward into the posttemporal fossa.

In *Acanthopsoides*, where the first epineural with a (primitive) anteromedial branch to the neural arch is on V6, the anteroventral tip of that epineural is bifid and extends downward toward the proximal part of the rib of V5, where it attaches to the epicentral ligament. The anteroventral tip of the epineural of V5 is expanded and bifid, and it is attached to the epicentral ligament of V5 close to its origin on the proximal part of the rib. The anteroventral tip of the epineural of V4 is much expanded and attaches (presumably by a ligament belonging to the epicentral series) to the posterodorsal surface of the capsule formed from the parapophyses of the fourth vertebra (Sawada, 1982, figs. 58, 62C), and the tip of the epineural of V3 is attached by ligament to the tip of the horizontal process from the fused second and third vertebrae (Sawada, 1982, figs. 58, 62C). In front of this there are three more strap-like epineurals (V1, V2, and one occipital) passing forward and attaching by ligaments to the occiput.

A fuller survey of the anterior intermusculars in otophysans is desirable, but features common to *Alestes* and *Acanthopsoides* include loss of the medial branch and the proximal attachment on the first four or five epineurals, and development of ligamentous attachments between tips of the anteroventral branches of the anterior epineurals and structures ventral to them; those structures include the swimbladder and transverse process of the third neural arch in *Alestes*, and the epicentral ligaments, capsule (fourth parapophyses), transverse process of the fused second and third vertebrae, and occiput in *Acanthopsoides*.

Some gonorynchiforms show similar modifications of the anterior intermusculars, but the situation is more complex than in otophysans because gonorynchiforms have epicentrals as well as epineurals on the anterior vertebrae. We describe the anterior intermusculars of *Gonorynchus* above, and those of *Chanos* and *Parakneria* are recorded in Table 4.

In *Chanos* (Rosen and Greenwood, 1970, figs. 3, 5, 6; Table 4), as in *Gonorynchus* and otophysans, epineurals are forked proximally (back to about V30 in *Chanos*, about V35 in *Gonorynchus*), and the anteromedial branch of the epineural is attached to the neural arch by ligament. The first anteromedial branch of an epineural is attached by ligament to the neural arch of V2. In front of this there are four more epineurals, each represented only by the anteroventral branch and posterior part of the bone; they have brush-like posterior ends and variously subdivided anterior ends. The first two of these pass forward into the posttemporal fossa. The epicentrals of *Chanos* are unusual in being widely forked distally (Rosen and Greenwood, 1970, fig. 3), a condition we have seen elsewhere only in kneriids (Lenglet, 1974; Table 4). They attach to the head of the rib on the anterior rib-bearing vertebrae and, by means of a long ligament, to the first two centra. There are two occipital epicentrals (the cranial ribs of Rosen and Greenwood, 1970), an enlarged one attached by ligament to the basioccipital, and one similarly attached to the exoccipital, just below the vagus foramen. Each of these has an anterodorsal branch, passing forward in the myoseptum into the rear of the posttemporal fossa. Both occipital epicentrals attach by ligament to the dorsal tip of the cleithrum distally. There is no Baudelot's ligament.

The intermuscular bones of kneriids were described by Lenglet (1974). Epineurals of our *Parakneria* (Table 4) are forked proximally from V7 back, and the anterior ones are attached by ligament low on the anterior margin of the expanded neural arches. Passing forward from V7, epineurals become more and more weakly ossified, and some of them are mere discontinuous slips of bone, remote from the neural arch, in the epineural ligament. On V3, the epineural ligament attaches to the centrum, just above the parapophysis, and contains a slip of bone distally. On V2, the right side of our specimen shows two ligaments originating together, low on the centrum, with a small slip of bone distally in the upper one, which we take to be the epineural, and the lower one leading to the epicentral. On the left side of the specimen there are two ligaments originating on the centrum, each leading to the head of one of the proximal branches of the epicentral; we assume that the epineural is fused here with the epicentral. There is no epineural on V1. Epicentrals of *Parakneria* are forked distally, like those of *Chanos* (cf. Lenglet, 1974, figs. 17–19; Rosen and Greenwood, 1970, fig. 3), and attach to the rib head on anterior rib-bearing vertebrae. On V2, the proximal end of the left epicentral is trifid, with an anterodorsal branch attaching to the centrum by a separate ligament, as noted above; we take this branch to be the incorporated epineural. The head of the epicentral of V1 is single and attaches to the centrum by ligament. There is one occipital epicentral (the cranial rib of Lenglet, 1974), a stout bone, forked distally, attaching to the exoccipital just behind the vagus foramen. *Kneria* has virtually the same configuration of the intermusculars as *Parakneria* (Lenglet, 1974). Among other gonorynchiforms, *Phractolaemus* (Thys van den Audenaerde, 1961) and *Cromeria* have the epicentral series represented only by a "cranial rib," and the diminutive *Grasseichthys* lacks epicentrals (and other intermuscular bones).

In *Gonorynchus* (described above; Figures 4–6), the anterior intermusculars are modified in ways comparable to those of kneriids. The attachment of the epineurals to the vertebra descends from the neural arch (V5 and successors) to the centrum (V3, 4). On V1–2 the epineurals are unattached. The anteroventral branches of an occipital epineural and of the epineurals of V1–3 fuse with the epicentral of the segment in front. There is no epicentral on V3; we have seen a comparable gap in an intermuscular series only in otophysans (e.g., no epineural on V3 in *Hemibarbus*, no epicentral on V4 in *Alestes*, no epicentral on V2–4 in *Hemibarbus* and *Hypentilium;* Table 4), in *Umbra* (no epicentral on V6–16; Table 4), and in some specimens of the aulopiform *Pseudotrichonotus* (no epineural on V3; Table 5).

To sum up, gonorynchiforms are shown to be monophyletic by the presence of epicentral bones (in addition to other features listed by Fink and Fink, 1981). Within gonorynchiforms, *Chanos* shows the most generalized condition, but it lacks medial branches on the three occipital epineurals and that of V1. *Gonorynchus* and kneriids show more extensive modifications of the anterior intermusculars, including associations between epineurals and epicentrals and descent of the epineural attachment on to the centrum. In otophysans, the anterior intermusculars are modified in connection with the Weberian apparatus. In our sample, the cyprinid *Hemibarbus* shows the most generalized condition, but both the epineural and epicentral series are interrupted in the region of the Weberian apparatus. The other genera have the anterior epineurals modified by loss of their medial branches, and in *Acanthopsoides* and *Alestes* the anteroventral branches of the anterior epineurals develop associations with structures ventral to them. One could sum up the generality common to otophysans and gonorynchiforms as a clearing away of intermuscular attachments to the dorsal parts of the first few vertebrae; in otophysans that clearing away is associated with development of the Weberian apparatus. The gonorynchiform condition is presumably a plesiomorphic version of the otophysan condition, as supposed on other grounds by Greenwood et al. (1966) and Rosen and Greenwood (1970).

ESOCOIDS.—In Table 4 we record the intermuscular bones and ligaments of four esocoids, two species of *Esox* and one each of Umbra and *Novumbra*. The relationships of esocoids are problematic. To date, the only proposal with any support (loss of a single character) is that esocoids are the sister group of all other euteleosts, including ostariophysans, all of which lack a fourth basibranchial toothplate, a structure present in esocoids (Fink, 1984). The intermusculars and anterior vertebrae may bear on this problem.

In *Umbra* the intermusculars are unremarkable, except for modification of the epicentral series. The anterior epineurals are simple; the first is in ligament in *U. krameri* and in bone in *U. pygmaea*. The posterior epipleurals and epineurals are unattached, and some of them are forked proximally. The epicentrals are all in ligament. *Umbra krameri* is unique among taxa we examined in having an extensive gap in the epicentral series, with ligaments from the occiput to V5 (the last one or two weaker than their predecessors) and then nothing until the series reappears at V17. In *U. pygmaea*, we found epicentral ligaments only on a few vertebrae in the middle of the trunk. In *Novumbra,* all epineurals and epipleurals are in ligament (they may ossify in larger specimens, Wilson and Veilleux, 1982), and we could find no epicentral ligaments.

In *Esox* (Figure 8) things are much more complex, including differences between *E. lucius* and *E. americanus* (respectively representing the species groups assigned to the subgenera *Esox* and *Kenoza* by Nelson, 1972). In *E. lucius,* the first four (or occasionally the first five, Figure 8) epineurals and in *E. americanus* the first two epineurals are broadly fused to the base of the neural arch. Both subgenera show epineurals attached to the centrum rather than the neural arch from about V10–25; this is an unusual specialization, which we record elsewhere only in *Xenomystus* (Table 3), *Anchoa* (Table 3), and in notosudid, scopelarchid, and evermanellid aulopiforms (Table 5) among lower (nonacanthomorph) teleosts. In *E. lucius,* epineurals are forked proximally from V7–8 onward and are attached to the neural arch or centrum by ligament (Figure 8A). In *E. americanus,* epineurals are like those of *E. lucius* from about V10 onward, but those of V3–8 are modified (Figure 8B). The epineurals of V3–6 are unbranched and consist only of the anteroventral limb and the posterodorsal body of the bone ($V_0$6, Figure 8B). The ligamentous connection to the neural arch remains on V3 and V7. The anteroventral tips of the epineurals of V3–8 are attached to the head of the rib of the same vertebra by means of the epicentral ligament (Figure 8B), and on V4–6, where the ligament to the neural arch is missing, this attachment to the epicentral ligament is the only attachment of the bone. *Esox* has only one anterior supraneural (Figure 8), an expanded, ovoid plate over the second and third neural arches (a series of small, unmodified supraneurals begins over about V10). In *E. americanus,* the second neural arch is swollen in the transverse plane and rostrocaudally expanded, extending back over the third centrum (Figure 8B); in *E. lucius,* the third (not the second) neural arch is similarly swollen and expanded (Figure 8A). The first rib is on V2 in all esocoids. In *E. americanus,* the first rib is coossified with the parapophysis in membrane bone, the ribs on V3–20 ossify in cartilage, and those on V21–31 ossify in membrane. In *E. lucius,* the first three ribs (on V2–4) are coossified with the parapophysis in membrane bone, and all but the last of the remaining ribs ossify in cartilage. These differences between *E. lucius* and *E. americanus* seem to be constant for the two subgenera; *E. masquinongy* and *E. reicherti* (subgenus *Esox*) show the *lucius* pattern, and *E. niger* (subgenus *Kenoza*) shows the *americanus* pattern.

When we first studied the intermusculars of *Esox,* it happened that we had access only to cleared-and-stained specimens with the *Kenoza* pattern (*E. americanus, E. niger*), showing an association between the anterior epineurals and the epicentrals. As Table 4 and the discussion in the preceding section indicate, gonorynchiforms (*Chanos* excepted) and otophysans are the only other teleosts in which we have found modifications of the intermusculars on the first few vertebrae that include association between the epineurals and the epicentrals (indicated by vertical lines in Table 4). We therefore jumped to the conclusion that the epineural/epicentral association was synapomorphous, and that in *Esox* we had found the sister group of the ostariophysans. We had to discard that idea for three reasons. First, species of the subgenus *Esox* (*E. lucius, E. niger, E. reicherti*), when they became available, proved to have no epineural/epicentral association, and there is good evidence that *Esox* is monophyletic (Nelson, 1972; Wilson, 1984). Second, there is no epineural/epicentral association in umbrids, and there is reasonable evidence that esocoids are monophyletic (Nelson, 1972; Rosen, 1974; Wilson, 1984). Third, *Chanos* and the cypriniforms we sampled have no epineural/epicentral association (Table 4), and there is good evidence that ostariophysans, gonorynchiforms, and otophysans are each monophyletic (Fink and Fink, 1981). The epineural/epicentral association therefore must be homoplastic in *Esox* (*Kenoza*), gonorynchoids, and otophysans. But we think it worth pointing out that beyond the epineural/epicentral association, *Esox* shares three other specializations of the anterior vertebrae with some or all ostariophysans.

The first of those is the presence of only one anterior supraneural, which is expanded (compared with the primitive rod-like or sigmoid supraneural) and is situated over the second and third neural arches, of which either the second (*Kenoza*) or third (*Esox*) is expanded. In kneriid gonorynchoids (including *Phractolaemus*) there is also a single expanded supraneural over the third neural arch (Thys van den Audenaerde, 1961; Lenglet, 1974), whereas in gonorynchids (including several Cretaceous forms) there is a series of four or five supraneurals over V2–5 or V2–6, with that over V2 or V3 enlarged in some Cretaceous genera (Gayet and Chardon, 1987). In otophysans there are one (characiphysans) or two (cypriniforms) enlarged supraneurals (the neural complex), those over V3 or V2–3 (Fink and Fink, 1981; Patterson, 1984). The derived condition of only one (*Esox,* kneriids, characiphysans) or two (cypriniforms) enlarged anterior supraneurals is certainly homoplastic in esocoids, gonorynchiforms, and otophysans because there are numerous unmodified supraneurals in umbrids (Wilson and Veilleux, 1982), *Chanos,* the Paleocene *Esox tiemani* (Wilson, 1984), and the Eocene otophysan *Chanoides,* which has unmodifed supraneurals over V5–10 in addition to the enlarged supraneurals forming the neural complex over V1–4 (Patterson, 1984).

The second anterior vertebral modification shared by *Esox* and ostariophysans is expansion of the anterior neural arches. Fink and Fink (1981:325, character 63) described the ostariophysan condition and stated that "the dorsomedial portions of the anterior neural arches are expanded and abut against each other and the posterior margin of the exoccipital, forming a roof over the neural canal. In other primitive teleosts, the neural arches are smaller and do not meet each other or the exoccipital." As Figure 8 shows, *Esox* shares this condition, with the difference that the accessory neural arch (ANA, Figure

8) is present, separating the first neural arch from the exoccipital. The anterior neural arches are not expanded in umbrids (Wilson and Veilleux, 1982, fig. 12), so that if esocoids are monophyletic, the derived state is homoplastic in *Esox* and ostariophysans.

The third anterior vertebral modification shared by *Esox* and some ostariophysans is coossification of the rib and parapophysis on V2 (*Kenoza*) or on V2–4 (*Esox*). In gonorynchiforms, the foremost rib (on V3) and parapophysis are not coossified (Greenwood et al., 1966; Rosen and Greenwood, 1970), but that coossification appears to be a necessary precondition for formation of the tripus (V3) and os supensorium (V4) in otophysans (Rosen and Greenwood, 1970; Fink and Fink, 1981; Patterson, 1984). Again, the esocoid and otophysan conditions must be homoplastic becaue the anterior ribs are unmodified in umbrids and gonorynchiforms.

We checked *Esox* and esocoids against Fink and Fink's (1981) list of fifteen ostariophysan characters. Esocoids share one of them (absence of basisphenoid, Fink and Fink's character no. 7); *Esox* shares two more (absence of supraneural in front of first neural arch, no. 58; expansion and abutment of anterior neural arches, no. 63, above and Figure 8); esocoids either lack the remaining ostariophysan characters (Fink and Fink's characters 8, 20, 54, 55, 57, 64, 111, 118), or those characters do not characterize ostariophysans (no. 41, as discussed in Patterson, 1984; no. 56).

Current opinion (Fink and Weitzman, 1982; Fink, 1984; Begle, 1992; Figure 9A) is that esocoids are the sister group of all other euteleosts, including ostariophysans. As noted above, the single (loss) character proposed in support of that grouping is absence of a toothplate on the fourth basibranchial (BB4) in all euteleosts except esocoids. In discussing this character, Fink and Weitzman (1982) reported that among clupeomorphs the toothplate occurs only in *Chirocentrus,* which would weaken it as primitive for clupeomorphs according to Grande's (1985) cladograms. However, Nelson (1970a) also recorded the toothplate in the pristigasteroid *Pellona* (but not in other pristigasteroids) and recorded fragmentary toothplates over BB4 in several engrauloids, rendering the assessment of the primitive state problematic. We note that the only ostariophysans with any basibranchial teeth or toothplates are *Gonorynchus* and characiforms; a toothplate over BB4 occurs only in certain characiforms, where reacquisition is inferred (Fink and Weitzman, 1982). In our view, the character is questionable. Fink and Weitzman (1982) and Fink (1984) placed the ostariophysans in a tetrachotomy, the other three branches of which are salmonoids, argentinoids + osmeroids, and neoteleosts (Fink and Weitzman, 1982), or argentinoids, osmeroids, and salmonoids + neoteleosts (Fink, 1984). Rosen (1985) placed ostariophysans in a trichotomy whose other two branches are argentinoids, and salmonoids + osmeroids + neoteleosts. We discuss the salmonoids, osmeroids, and argentinoids below, and we argue that they form a monophyletic group within which salmonoids and osmeroids are sister taxa (Figure 9B). That grouping simplifies Fink and Weitzman's (1982) and Fink's (1984) tetrachotomies and modifies Rosen's (1985) trichotomy; the result is a trichotomy whose three branches are ostariophysans, salmoniforms (argentinoids, salmonoids + osmeroids), and neoteleosts. We have suggested (Johnson and Patterson, 1993:600) that this trichotomy may be resolved by the median caudal cartilages (CMC) of Fujita (1990, table 3). CMC are unknown in osteoglossomorphs, elopomorphs, clupeomorphs, esocoids, and ostariophysans (except for a cartilage in the position of a lower CMC in Fujita's specimen of the cyprinid *Tanakia*). CMC are present in argentinoids, osmeroids, salmonoids, and in stomiiform, aulopiform, and myctophiform neoteleosts; they are absent (secondarily) in acanthomorphs (Johnson and Patterson, 1993). We regard CMC as synapomorphous for Euteleostei minus esocoids and ostariophysans, a group for which we here (Figure 9) adopt the name Neognathi (Rosen, 1973; Rosen's usage differed from ours by including esocoids).

Within the Clupeocephala, there are thus four monophyletic groups: (1) Clupeomorpha, (2) Ostariophysi, (3) Esocoidei, (4)

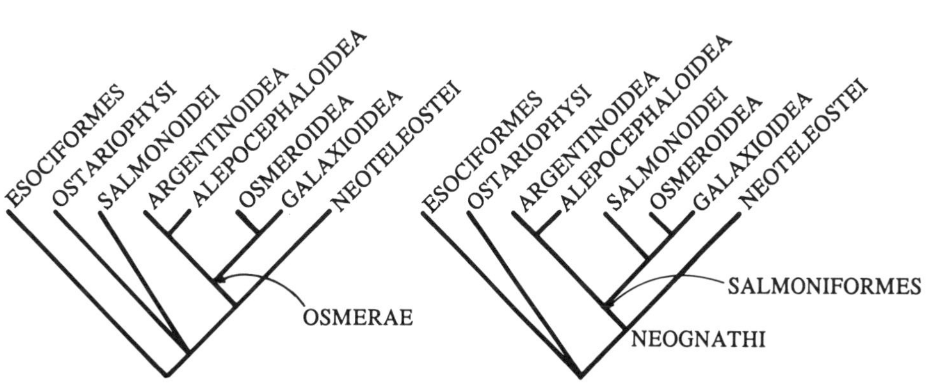

FIGURE 9.—Cladograms contrasting Begle's (1991, 1992) view of the interrelationships of lower euteleosts (left) with that proposed here (right).

Neognathi. We know of no morphological character that unambiguously groups any two of these four. The adipose fin, nuptial tubercles, and a stegural have been used to group (2) + (3) + (4) (Patterson and Rosen, 1977; Lauder and Liem, 1983), but the adipose and nuptial tubercles do not occur in esocoids, and their stegural is as dubious as that of ostariophysans. Molecular evidence from nucleotide sequences of partial large subunit ribosomal RNA (Le et al., 1993) and ependymins (Müller-Schmid et al., 1993) favors grouping (1) + (2), but esocoids are not yet sampled for ribosomal RNA, and the sampling in ependymins also is poor (*Clupea, Esox, Oncorhynchus,* and two cyprinids). Although we have found no convincing evidence that ostariophysans (2) and esocoids (3) belong together, and are unwilling to appeal to "underlying synapomorphies" (Saether, 1983), we believe that the anterior vertebral modifications unique to some or all esocids and some or all ostariophysans are sufficiently striking to encourage further search for ostariophysan/esocoid synapomorphies.

SALMONOIDS, OSMEROIDS, AND ARGENTINOIDS.—Salmonoids and osmeroids share cartilaginous rods in the epicentral ligaments (see p. 14 and Table 4, Plate 1D,H), and, as noted above, the distribution of those structures in both groups implies that they are primitive for each according to Sanford's (1990) cladogram of salmonoids and Begle's (1991) cladogram of osmeroids. Salmonoids and osmeroids also share the absence of ossified epipleurals; the osmerid *Spirinchus* (pers. obs.) and the galaxiines *Galaxias* and *Paragalaxias* (McDowall, 1969, 1978) are exceptions, but *Spirinchus* and the galaxiines are each among the most derived members of their clades (Begle, 1991; Wilson, 1992), and, as each has several successive outgroups that lack epipleurals, their epipleural bones must be interpreted as reversals. Although we infer that absence of epipleural bones is primitive for teleosts, the universality of a long series of epipleural bones in elopomorphs, clupeomorphs, esocoids, ostariophysans, argentinoids, stomiiforms, aulopiforms, and myctophiforms (Tables 3–5) shows that their absence in salmonoids and osmeroids is secondary or derived and so potentially synapomorphous. Thus, there are two intermuscular characters, presence of epicentral cartilages and absence of epipleural bones, implying that salmonoids and osmeroids are related.

At present, the relationships of salmonoids are unresolved (Fink and Weitzman, 1982; Fink, 1984; Rosen, 1985; Sanford, 1990; Begle, 1992). A sister-group relationship between osmeroids and argentinoids (Argentinoidea + Alepocephaloidea; Greenwood and Rosen, 1971; Begle, 1991, 1992) was indicated in Fink and Weitzman's (1982, fig. 23) cladogram, but no support was given for that node (cf. Fink, 1984). However, Begle (1991, 1992) cited characters supporting the monophyly of his Osmerae (Argentinoidei + Osmeroidei; Figure 9A). Argentinoids lack epicentral cartilages, so far as we know, and have epipleural bones. Thus, there are two sets of contradictory characters bearing on the relationships of osmeroids: the features listed by Begle as relating them to argentinoids and the two features of the intermusculars (cartilaginous epicentrals, no epipleurals) indicating that they are immediately related to salmonoids. Because of this conflict, we evaluate the characters proposed by Begle (1991, 1992; his character numbers are given in square brackets).

In his 1991 paper, Begle (figs. 2, 3) cited four characters relating osmeroids and argentinoids: (1) "loss" of nuptial tubercles [33]; (2, 3) "loss" of basisphenoid [57] and orbitosphenoid [58]; and (4) presence of a ventral cartilaginous vane on the first basibranchial [64]. In a second paper, Begle (1992) diagnosed the Osmerae (Argentinoidei + Osmeroidei) by characters 2 (rephrased as "reduction/loss of basisphenoid" [57]) and 4 [64] in the above list, and by a new one: (5) fusion of rudimentary neural arches to ural centra [5]. Begle's (1992:353) character 1 [33] in the above list appears as an "additional character... that may occur at this node under alternative optimization schemes," together with: (6) loss of endopterygoid teeth [3]; (7) reduced pterosphenoids [63]; and (8) basihyal teeth [106] (four states are found in osmeroids and argentinoids, Begle, 1992:363). In addition, Begle (1992) cited two further characters uniting Osmerae and Neoteleostei, which therefore distinguish Osmerae from salmonoids: (9) extended alveolar process of premaxilla [83]; and (10) reduction in laminar bone on anterior margin of hyomandibular [100]. Evaluation or criticism of these 10 characters follows; asterisks indicate characters listed as diagnostic by Begle (1992), with the implication that he acknowledged ambiguity in the distribution of characters without asterisks.

1. *Absence of nuptial tubercles.* As Begle (1991:39) noted, nuptial tubercles occur among osmeroids in Osmeridae, *Lovettia, Stokellia,* and *Retropinna.* The character is ambiguous.

2, *3. *Absence of orbitosphenoid and basisphenoid.* Both of these bones occur in argentinoids. They were described in alepocephaloids by Gosline (1969:198, with references to Gegenbaur's earlier work) and in argentinids by Chapman (1942), and we have confirmed that both bones are well developed in dried skeletons of *Argentina* and *Alepocephalus.* Begle (1992:362) also acknowledged that the basisphenoid and orbitosphenoid are present in a range of argentinoids. In our view the characters are mistaken.

*4. *Ventral cartilaginous vane on first basibranchial.* Begle (1991, fig. 5) explained this character by means of drawings of the first basibranchial (B1) in the salmonoid *Thymallus,* the osmerid *Thaleichthys,* and the platytroctid *Sagamichthys. Thymallus* has a small, cuboid, ossified B1 (Begle's "unmodified" state), whereas the other two genera have a larger, cartilaginous B1 with a ventral vane or keel (Begle's "modified" state). There are several conditions of B1 in osmeroids and argentinoids. Begle (1991, fig. 3) noted that B1 shows the unmodified state in *Lepidogalaxias* and more-derived galaxioids, and we find that it also is unmodified in the argentinoids *Glossanodon, Argentina,* and *Opisthoproctus* (cartilaginous in the first, ossified in the other two), whereas we find in the osmerids *Allosmerus* and *Plecoglossus* it has the modified shape but is ossified. The character is ambiguous.

5. *Reduction or loss of endopterygoid teeth.* Osmeroids (except salangids and *Lepidogalaxias*) have a band of enlarged teeth along the medial margin of the endopterygoid, whereas argentinoids, salangids, and *Lepidogalaxias* have the endopterygoid toothless. We see two characters here: absence (versus presence) of teeth and modification (versus uniformity) of dentition.

*6. *Fusion of rudimentary neural arches to ural centra.* In his 1991 paper (page 48), Begle treated this as a two-state character, with the derived state ("uroneural 1 fuses to compound centrum, followed in some forms by fusion with neural arches") found only in Osmeroidea. In the 1992 paper (page 361), it became a three-state character, with state 1 (recorded in argentinoids, most alepocephaloids, and in southern osmeroids) as "rudimentary neural arches fuse to centrum, followed in some cases by fusion to first uroneural," and state 2 (recorded in northern osmeroids) as "rudimentary neural arches fuse first to uroneural, followed in some cases by fusion to centrum" ("followed" is used here with reference to ontogeny, Fink and Weitzman, 1982:83; Begle, 1992:356). This character originated with Patterson (1970), who used the alternative derived states to distinguish argentinoids, alepocephaloids, and galaxiids (state 1, distinct neural arch on PU1 and often on U1) from northern osmeroids, *Prototroctes,* and *Retropinna* (state 2, no neural arch on PU1 or U1). Begle (1992:356) writes that "fusion of the rudimentary neural arches to the centrum is unambiguously the primitive condition for the Osmerae," implying that state 2 is a modification of state 1. But this follows only if the Osmerae are demonstrably monophyletic. Given that retropinnids (Rosen, 1974, fig. 27), *Lovettia* (Rosen, 1974, fig. 19), and salangids (Rosen, 1974, fig. 26; Roberts, 1984) show the northern osmeroid pattern, with no caudal neural arches (state 2), Begle's (1991) cladogram indicates that the caudal neural arches (state 1, supposedly primitive) in galaxiines and *Aplochiton* are nonhomologous with those in argentinoids.

Rudimentary caudal neural arches (assignable to PU1 or U1) do not occur in neoteleosts (Fujita, 1990), and in salmonoids they occur only occasionally (as individual variants, commonest in *Prosopium* and *Thymallus;* Arratia and Schultze, 1992).

7. *Reduced pterosphenoids.* The reduced state is characterized by Begle (1992:362) as "widely separated, not meeting at midline." In our view, the latter is the primitive state. The pterosphenoids are widely separated and fail to meet in the midline in Jurassic pholidophorids and leptolepids (Patterson, 1975, figs. 63, 72, 79, 89), and in *Hiodon, Elops,* esocoids, salmonids, etc.

8. *Basihyal teeth.* Begle (1992) acknowledged that this feature is ambiguous, with state 1 (marginal fangs) occurring in salmonids and osmeroids, and state 4 (teeth absent) in most argentinoids.

*9. *Extended alveolar process of premaxilla.* Begle (1991, 1992) illustrated this character by comparing the pattern in osmeroids, argentinoids, and a stomiatoid with that in salmonids, where the premaxilla and maxilla are in tandem, with no extension of the premaxilla beneath the maxilla; the salmonid pattern was taken to be primitive. As with number 6 above, we regard the polarity as inverted here because there is an extended alveolar process in (for example) *Hiodon, Elops, Megalops, Albula,* most clupeoids, all esocoids, etc.

*10. *Reduction in laminar bone on anterior margin of hyomandibular.* This is a difficult character to evaluate because it is one of degree. Begle (1992) noted that laminar bone is present among alepocephaloids in *Leptochilichthys* and *Narcetes* (which occupy two of the three basal nodes in his cladogram of the group) and among osmeroids in *Plecoglossus* and *Prototroctes* (both derived members of the group in his cladogram). Laminar bone also is well developed in *Alepocephalus* (Gosline, 1969, fig. 5 and a dried skeleton). The problem of evaluating the character is exemplified by Sanford (1988:134), who judged that an anterior laminar extension of the hyomandibular is present in all argentinoids, and he wrote that "this mosaic distribution indicates that an anterior laminar extension ... has arisen, and perhaps been lost, many times amongst teleosts."

From this survey, we conclude that there is, as yet, no unequivocal support for argentinoid-osmeroid relationships, and we offer the alternative that salmonoids are the sister group of osmeroids (Figure 9B) because of the presence in both of cartilaginous epicentrals and the absence in both of ossified epipleurals.

Within argentinoids, the intermusculars suggest one synapomorphy corroborating the sister-group relationship between alepocephaloids and argentinoids: the tips of the first three (*Glossanodon, Leptoderma, Searsia*) or four (*Argentina*) epineurals are displaced ventrally relative to their successors. The intermusculars also provide a character corroborating the monophyly of alepocephaloids: the epipleurals extend unusually far forward, to V3 (recorded in *Searsia* in Table 4, and observed also in *Bathytroctes, Leptochilichthys, Rinoctes, Talismania*). Developmental implications of this feature are mentioned below in the section on aulopiforms.

Salmoniformes: Salmoniformes as circumscribed by Greenwood et al. (1966) now are seen to be nonmonophyletic (e.g., Rosen 1974; Fink, 1984). But after removal of the esocoids, stomiatoids (= Stomiiformes), and myctophoids (= Aulopiformes and Myctophiformes), there remain three groups: salmonoids (Sanford, 1990), argentinoids (Argentinoidea + Alepocephaloidea; Greenwood and Rosen, 1971; Begle 1992), and osmeroids (Osmeroidea + Galaxioidea; Begle, 1991). Sanford (1990:149) proposed two characters indicating that these three form a monophyletic group: absence of radii on the scales (a character that we have not tried to evaluate) and well-developed marginal teeth on the basihyal (a character discussed by Nelson, 1970b, Fink and Weitzman, 1982, and Begle, 1992). Jamieson's (1991) review of spermatozoan structure provided one further salmoniform synapomorphy, a single, annular mitochondrion, recorded in two alepocephaloids (Argentinoidei), in salmonids, and in *Galaxias* (Osmeroidei); however, the mitochondria are not annular in *Lepidogalaxias*. The intermusculars provide one further hint that these three groups may be related.

In elopomorphs, clupeomorphs, and ostariophysans, the posterior epineurals and epipleurals are forked proximally (Tables 3, 4). Among esocoids, *Esox* has the epineurals forked proximally but not the epipleurals (Table 4). Argentinoids, salmonoids, and osmeroids are unique among lower (non-neoteleost) elopocephalans in lacking proximal forking of epineurals and epipleurals, and this might be a synapomorphy of Salmoniformes. The character is weakened because proximal forking of epineurals and epipleurals also is lacking in stomiiforms (Table 5; Weitzman, 1974; Fink and Weitzman, 1982) and is unusual in aulopiforms (recorded only in *Chlorophthalmus* and *Parasudis*, Table 5). It is general in myctophiforms (Table 5).

If the Salmoniformes are monophyletic, their intermusculars indicate that argentinoids are the most primitive members of the group. They retain an extensive series of epineurals fused to the neural arches (V1-27 in *Alepocephalus*, V1-23 in *Searsia*, V1-21 in *Argentina*, V1-20 in *Glossanodon*), lack epicentral cartilages, and have epipleural bones.

Lepidogalaxias: The relationships of *Lepidogalaxias* have long been a problem (Rosen, 1974; Fink and Weitzman, 1982; Fink, 1984). Begle (1991) placed *Lepidogalaxias* among galaxioid osmeroids, as the sister group of salangids + galaxiids. Although salangids lack ossified intermusculars (*Salangichthys*, Table 4; Roberts, 1984, fig. 1), the intermusculars of *Lepidogalaxias* give general support to this proposal. *Lepidogalaxias* (Table 4) shows several unique features in the intermusculars: there are no epineurals or epipleurals in either bone or ligament, and the only intermusculars are epicentrals, which are represented by a series of ligaments but with those on V1-6 containing bone distally. These bones are remote from the vertebrae and are forked distally, just beneath the skin. Distal forking of the epicentral bones occurs elsewhere only in *Chanos* and kneriids, so far as we know, and as argued above (p. 14), we believe that the bones in *Lepidogalaxias* are cartilage bones, ossifications of the epicentral cartilages present in other osmeroids. Apart from *Lepidogalaxias*, the only other genus that lacks epineural bones in Tables 3-5 is the retropinnid *Stokellia*, where epineurals are represented by ligament (*Retropinna* and *Salangichthys* show the same condition). The ribs of *Lepidogalaxias* are all in membrane bone, another unusual (although not unique) condition in non-neoteleosts, and the first rib is on V2, as in *Galaxias*.

Neoteleosts and Eurypterygians: We found no features of the intermusculars bearing on monophyly of the Neoteleostei (Stomiiformes + Aulopiformes + Myctophiformes + Acanthomorpha) or Eurypterygii (Neoteleostei minus Stomiiformes). Characters corroborating the monophyly of these groups recently were enumerated by Johnson (1992).

We had hoped that the intermusculars might provide some information about the relationships of ateleopodids, which were placed in an unresolved trichotomy with stomiiforms and eurypterygians by Olney et al. (1993). Unfortunately, our observations on two cleared-and-stained specimens of *Ateleopus* indicate that ateleopodids lack all intermuscular bones and ligaments, and they also lack ribs and Baudelot's ligament. We have not included comments on absence in ateleopodids in our sections discussing the distribution of components of the intermuscular system and associated structures.

AULOPIFORMS.—We record the intermusculars of 18 aulopiforms in Table 5, a wide sample illustrating the remarkable variations within this group and their potential value in systematics. Our sample covers members of 13 of the 15 nominal families (Okiyama, 1984; Johnson, 1992) or families and subfamilies (Nelson, 1984) currently recognised in the Aulopiformes, i.e., all except the monotypic Anotopteridae (included in Paralepididae by Johnson, 1992) and Omosudidae. Aspects of the intermuscular bones in *Anotopterus* and *Omosudis* are illustrated by Rofen (1966a, 1966b). *Omosudis* is held to be closely related to *Alepisaurus* by Johnson (1992); we comment on the structure and relationships of *Omosudis* below on the basis of a small cleared-and-stained *O. lowei* (USNM 219982, 81 mm SL), which was not decipherable in sufficient detail for entry in Table 5.

Aulopiforms are uniquely characterized by having epipleural bones extending far forward, always to V2 and frequently to V1 (Table 5). The only other groups with epipleurals extending forward to the anterior vertebrae are alepocephaloids (to V3, recorded in *Searsia* in Table 4, and observed also in *Bathytroctes, Binghamichthys, Leptochilichthys, Rinoctes*, and *Talismania*) and clupeoids (Table 3), and, in the latter, the anterior epipleurals are unattached to the axial skeleton (notation D_0 or B_0 in Table 3). In aulopiforms, well-developed, attached epipleurals extend forward to V1 or V2. This specialization forms the basis for other remarkable modifications within the group. Although we have little evidence bearing on the ontogeny of epipleural bones in aulopiforms, their mode of development evidently differs from that observed or inferred in other teleosts, where epipleurals develop rostrally and caudally from the region of the first caudal vertebra. In larval *Saurida,* epipleurals ossify caudally from V3, with that on V2 ossifying later than that on V3; the pattern in *Aulopus* (Table 5), with a small, unattached epipleural on V2, implies that development is as in *Saurida.* In other aulopiforms, the pattern of epipleurals in *Ahliesaurus, Coccorella, Paralepis,* and *Alepisaurus* (Table 5) is such that the bones can have developed only in a rostrocaudal gradient; in *Ahliesaurus* and *Coccorella* epipleural bones are almost or completely confined to the abdominal region, with a long series of ligaments extending into the caudal region, and in *Paralepis* and *Alepisaurus* the epipleurals on V1-2 are fused to the centrum, and the series passes back either into a long series of ligaments (*Alepisaurus*) or a long series of unattached bones (*Paralepis*). This change in mode of development may be due to associations between anterior epipleurals and the horizontal septum and its contained epicentrals that occur in various aulopiforms. These associations are discussed below.

In alepocephaloids, the only other group with attached epipleurals extending almost as far forward as in aulopiforms (V3 vs. V1-2), we have no information on development; however, because in *Searsia* (Table 4) the epipleural series is confined to the abdominal region (the last bone is attached by ligament to the first caudal vertebra), it is evident that the series does not develop from the primitive focus, at about the level of the first caudal vertebra. In *Rinoctes* the pattern is similar, with the last epipleural bone attached by ligament to the first haemal spine and about 10 ligaments, without included bone, beyond it. It may be that some or all alepocephaloids have altered the mode of development of epipleurals in a way comparable (although obviously nonhomologous) with that in aulopiforms.

Bathypterois (Ipnopidae) shows what seems to be the most generalised pattern of intermusculars in our sample of aulopiforms, with long, coextensive series of epineural (V1 to ~V50) and epipleural (V2 to ~V50) bones and epicentral ligaments. *Bathypterois* lacks the accessory neural arch, a structure present in *Aulopus* (Aulopidae), *Harpadon, Saurida* (both Harpadontidae), and *Synodus* and *Trachinocephalus* (both Synodontidae). *Aulopus,* generally taken to be the most primitive aulopiform, has a significant modification of the epicentral ligaments and epipleural bones in the abdominal region: from about V5 back to about V20 epicentral ligaments originate not on the centrum but on the epipleural bone (notation L_E in Table 5), near its base anteriorly and posteriorly but at up to about one-third of the length of the epipleural on about V10-15. The implication of this association is that the proximal part of these epipleurals is in the horizontal septum, not below it as is normal.

The larval *Scopelarchoides* (Scopelarchidae) illustrated in Figure 10 illuminates the condition in *Aulopus* and several other aulopiforms. In this *Scopelarchoides* (35 mm SL) all the intermusculars are still in ligament. On V1-11 there are only two ligaments on each vertebra, the epineural, originating on the neural arch (V1-3) or centrum (V4-11), and one that originates on the parapophysis and lies in the horizontal septum. On V7 the lower ligament bifurcates near its tip, and on V8-11 the bifurcation is successively closer to the centrum. From V12 onward there are two lower ligaments on each vertebra, an epicentral in the horizontal septum and an epipleural below it. In adult *Scopelarchoides* (Table 5) there are epipleural bones forward to V1, and those on V2-10 have an epicentral ligament closely applied to their upper surface. Both bone and ligament lie in the horizontal septum, but posteriorly (beyond V10) the bones and ligaments diverge, with the epicentrals moving up on to the centrum and the bones down on to the haemal arch. Here, ontogeny shows that the epicentral and epipleural series have coalesced anteriorly, with the anterior epipleurals displaced dorsally into the horizontal septum; the situation is the inverse of that in beryciform acanthomorphs (discussed below), where the anterior epineurals are displaced ventrally into the horizontal septum. In Table 5, epipleural bones lying in the horizontal septum are indicated in bold type (e.g., $\mathbf{B}_3$ for one originating on the parapophysis),

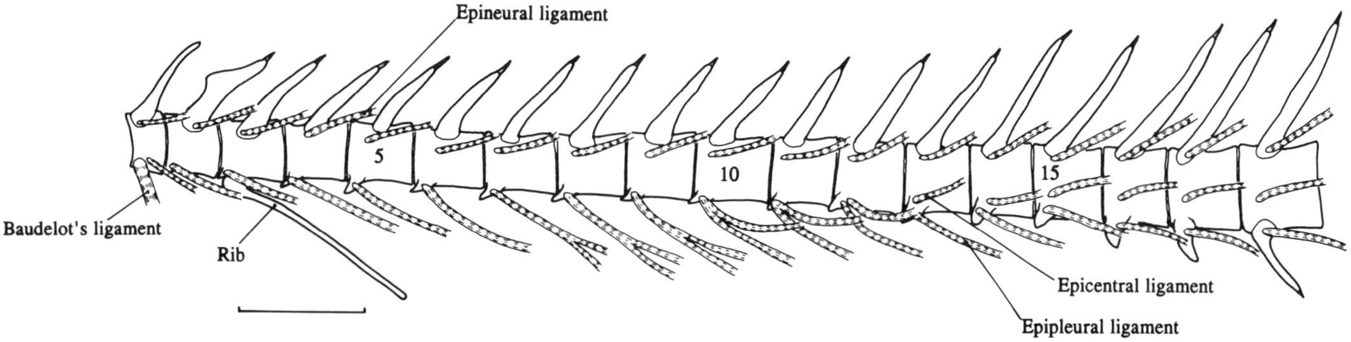

FIGURE 10.—Anterior vertebrae and associated intermuscular ligaments in larval *Scopelarchoides signifer* Johnson, USNM 274385, 35 mm SL. At this stage, the only ossified structures are centra (including the parapophyses) and tips of neural and haemal spines. All intermusculars are ligaments, and the one rib visible is the first, on V3, which is the only one ossifying in cartilage. Every fifth vertebra is numbered. (Scale bar 1 mm.)

just as in Tables 7 and 8 bold type indicates epineurals in the horizontal septum.

In the notosudid *Scopelosaurus* (Table 5), the epipleurals on V2–24 are in the horizontal septum, and the region where they turn ventrally and leave the septum is indicated by distal bifurcation of the bones on V20–24, with the upper branch lying in the horizontal septum and the lower branch below it. The first epicentral ligament occurs on V20, together with the first bifurcate epipleural, and its distal part is applied to the upper branch of the epipleural. Posteriorly, behind V24, epicentrals and epipleurals diverge, with the epicentral ligaments moving up on to the centrum and the epipleurals down on to the haemal arch. In *Ahliesaurus*, the other notosudid that we sample in Table 5, there is distal bifurcation of a single epipleural, the last one in the horizontal septum (on V19); epicentral ligaments first appear four vertebrae in front of this, as in *Scopelosaurus*, but the five ligaments that overlie epipleurals in the horizontal septum are closely applied to the upper margin of the bones. In the evermanellid *Coccorella* (Table 5), the epipleurals on V1–12 lie in the horizontal septum, and epicentral ligaments are applied to the upper surface of the bones on V6–12. The notosudids, scopelarchid, and evermanellid in our sample share epineural bones whose origin descends from the neural arch to the centrum (and even to the parapophysis in *Scopelosaurus;* also in *Scopelarchus* (USNM 234988) and *Evermanella* (Rofen, 1966c, fig. 185)) from about V5 back to at least V15.

According to Johnson et al. (1989), morphological evidence not including the intermusculars indicates that *Pseudotrichonotus* is the sister taxon of the Synodontoidea, represented in Table 5 by the synodontids *Synodus* and *Trachinocephalus* and the harpadontids *Harpadon* and *Saurida.* These genera agree in having the anterior epipleurals in the horizontal septum, on about V2–10 in *Harpadon,* V2–20 in *Pseudotrichonotus,* V2–25 in *Trachinocephalus* and *Saurida,* and V2–35 in *Synodus. Pseudotrichonotus, Synodus,* and *Trachinocephalus* also agree, and differ from other aulopiforms, in having an abrupt and obvious transition between epipleurals in the horizontal septum and below it. In other aulopiforms with dorsally displaced epipleurals, such as the notosudids, scopelarchid, and evermanellid discussed above, there is a gradual transition between the bones in and below the septum, typified by the successively more deeply forked ligaments on V8–11 in larval *Scopelarchoides* (Figure 10) and the successively more deeply forked bones on V20–24 in *Scopelosaurus.* In *Pseudotrichonotus, Synodus,* and *Trachinocephalus* the point where the epipleurals leave the horizontal septum is marked by an abrupt transition, with a stout, posterolaterally directed bone closely associated with an epicentral ligament on one vertebra, and a slender, ventrolaterally inclined bone with a free epicentral ligament above it on the next. Epicentral ligaments first appear above the last three or four dorsally displaced epipleurals in *Pseudotrichonotus* and *Trachinocephalus,* above the last six or seven in *Harpadon* and *Saurida,* and above about the last 10 in *Synodus.* In *Synodus,* these first 10 epicentral ligaments are short and insert on the upper surface of the proximal part of the epipleural bone, so that the bone appears to be slung from the centrum by a short ligament. The same pattern occurs in *Saurida, Trachinocephalus,* and *Pseudotrichonotus* in the epicentral ligaments overlying dorsally displaced epipleurals (notation $_{p}L$ in Table 5). In some specimens of *Synodus* the first free epicentral ligament (above the first ventrolaterally inclined epipleural) is ossified so that there is a single epicentral bone, but this condition is not constant, even on both sides of a specimen.

As noted above, Johnson et al. (1989) treated *Aulopus* as the sister group of *Pseudotrichonotus* + synodontoids. Synodontoids share with *Aulopus* the accessory neural arch (absent in *Pseudotrichonotus*) and ventral displacement of the distal part of the first one to three epineurals (notation b_l in Table 5). *Aulopus* has the first three epineurals modified in this way, *Pseudotrichonotus* has the first two modified and the third absent (Table 4) or the first three modified (ZUMT 55678), *Synodus* has the first two or three modified, *Trachinocephalus* the first, and in *Saurida* the first is modified in *S. brasiliensis* and *S. suspicia* but not in *S. gracilis* nor *S. normani* (the species recorded in Table 5). The first epineural is unmodified in *Harpadon.*

Other aulopiforms sampled in Table 5 (apart from the ipnopid, aulopid, synodontoids, notosudids, scopelarchid, and evermanellid discussed above) are the chlorophthalmids *Chlorophthalmus* and *Parasudis,* the paralepids *Paralepis* and *Macroparalepis,* and *Alepisaurus, Bathysaurus,* and *Gigantura,* each representing a monotypic family. In *Chlorophthalmus* and *Parasudis,* many epineurals and epipleurals are forked proximally, and the posterior epineurals and epipleurals lack the anteromedial fork and so are unattached to the axial skeleton (notation V_0 and D_0 in Table 5). This is a pattern, common in lower elopocephalans (Tables 3, 4), that we have not observed elsewhere in aulopiforms. But in other respects *Chlorophthalmus* and *Parasudis* are very different. *Chlorophthalmus* has the first two epineurals modified in the same way as *Aulopus* and most synodontoids (notation b_l in Table 5), and it has the first five or six epipleurals displaced dorsally into the horizontal septum. In other aulopiforms showing this displacement, the epicentral series of ligaments begins above the last epipleurals in the horizontal septum, but in *Chlorophthalmus* the first obvious epicentral ligament occurs at about V15, well behind the last dorsally displaced epipleural (in some specimens we saw indications of very small ligaments inserting on epipleurals anterior to this point, suggesting the pattern in *Aulopus,* but after examining a series of cleared-and-stained specimens of more than one species we were unable to convince ourselves that the ligaments were real). In contrast, *Parasudis* has an extensive series of epicentral bones from the occiput back to about V30 (continued caudally by a few ligaments), has no epipleurals in the horizontal septum, and has the first epineurals unmodified. Thus, the pattern in *Parasudis* most closely resembles the generalized pattern in *Bathypterois*, with the difference that the epicentrals are ossified.

Epicentral bones also occur in *Alepisaurus* (and *Omosudis*), where there is a long series from V3 onward. *Alepisaurus*, *Paralepis*, and *Omosudis* agree in having the first rib on V1, in having the rib and epipleural of V1 and V2 fused to the centrum, and in having most epineurals unattached to the axial skeleton; in *Paralepis* and *Omosudis*, most epipleurals also are unattached, whereas they are normal in *Alepisaurus*. The paralepid *Macroparalepis* differs from *Paralepis* in having only a few epineurals (in bone on V1–4 and in ligament on V5-7) and epipleurals (V1–5) and in having the first rib on V5, but it shares with *Paralepis* a fused epineural, one free epipleural (many in *Paralepis*), and Baudelot's ligament attaching broadly to V1 and V2.

Gigantura is unique in having many epineurals and epipleurals bifid proximally (a few epineurals are trifid), with both branches attaching to the centrum, and in having no ribs; only the first epipleural appears to be in the horizontal septum. Finally, in *Bathysaurus*, which is conventionally allied with or included in the Synodontidae, the pattern is as strange as in *Gigantura*, with which *Bathysaurus* shares absence of epicentral bones or ligaments. In *Bathysaurus*, the first few epipleurals (V1–8) are in the horizontal septum, almost all the epineurals originate on the centrum (rather than the neural arch) and are fused to it, and all but the first of the ribs originate not on the parapophyses but by ligament on the epipleurals (notation $L\underline{B}_E$ in Table 5).

Table 6 shows a matrix for 26 characters, all but one drawn from the data on aulopiforms in Table 5, and Figure 11 shows cladograms derived from that matrix (see caption to Figure 11 for procedures). Although we do not pretend that Figure 11 contains the truth on aulopiform interrelationships, we contend that it may be nearer the truth than other schemes currently available (Johnson, 1982; Okiyama, 1984; Hartel and Stiassny, 1986). In Johnson's (1982, fig. 20) scheme, *Aulopus* is remote from all other aulopiforms; notosudids, scopelarchids, chlorophthalmids, and ipnopids are placed with myctophiforms in the pattern [[Myctophidae, Neoscopelidae] [Notosudidae [Scopelarchidae, Chlorophthalmidae, Ipnopidae]]]; the remaining aulopiform families are placed in a third group in the pattern [[Bathysauridae [Synodontidae, Harpadontidae]] [[Paralepididae, Anotopteridae] [Evermanellidae [Omosudidae, Alepisauridae]]]]; and giganturids are not included. In Okiyama's (1984, table 69) phenetic scheme, aulopiforms fall into four groups: Aulopidae; Chlorophthalmidae + Ipnopidae + Notosudidae + Scopelarchidae; Bathysauridae + Harpadontidae + Synodontidae; and Alepisauridae + Anotopteridae + Evermanellidae + Omosudidae + Paralepididae. Okiyama's groups are, therefore, the same as the four major aulopiform components of Johnson's scheme, and again, giganturids are not included. According to Hartel and Stiassny (1986), *Parasudis* is the sister taxon of *Chlorophthalmus*, those two genera comprising the Chlorophthalmidae (Hartel and Stiassny transferred *Bathysauropsis* from Chlorophthalmidae to Ipnopidae). *Parasudis* differs from *Chlorophthalmus* and most other aulopiforms in having a full series of epicentral bones, including (in our specimen) one on the occiput and one on the notochordal sheath in the "cervical gap" of Rosen (1985) (Hartel and Stiassny, 1986, fig. 6A, illustrated a *Parasudis truculentus* in which there is no occipital epicentral and the first bone originates on the centrum of V1). Aulopiforms were nonmonophyletic for Hartel and Stiassny (1986), as they were for Rosen (1985). That notion is contradicted by the aulopiform synapomorphies listed by Johnson (1992) and by the unique pattern of epipleurals in the group. As for Hartel and Stiassny's (1986) Chlorophthalmidae (*Chlorophthalmus* and *Parasudis*), evidence from the intermusculars indicates that they might be monophyletic (Figure 11B, monophyly supported only by proximally forked epineurals and epipleurals, with the posterior members of each series unattached) or nonmonophyletic (Figure 11C). The link between *Chlorophthalmus* and synodontoids in Figure 11C is very weak (supported only by descent of the anterior epineurals), but in that tree a number of more substantial characters distinguish *Chlorophthalmus* from *Parasudis*.

The relationships of *Gigantura* are unknown. Giganturids were not included in the Aulopiformes by Okiyama (1984), and R.K. Johnson's (1984; Johnson and Bertelsen, 1991:27) comments suggest that they might be the sister taxon of aulopiforms. The dorsal gill arches of giganturids are greatly reduced (Rosen, 1973) and do not show the features used by Johnson (1992) to characterize aulopiforms. Nevertheless, the intermusculars indicate that *Gigantura* belongs within aulopiforms, specifically with alepisauroids, and that it may be linked with *Bathysaurus*, usually placed in the synodontoids.

Within alepisauroids, the implication of paralepidid paraphyly in Figure 11 may be artifactual and due only to reduction of the intermusculars in *Macroparalepis*.

We hope that Figure 11 may be enough to spur those interested in showing our errors to produce better schemes based on other aspects of aulopiform morphology.

Ctenosquamates: The Ctenosquamata comprise the Myctophiformes and Acanthomorpha (Johnson, 1992). Two features of the intermusculars are consistent with ctenosquamate monophyly, although both are subject to so much homoplasy in other teleostean groups that they can hardly be used as characters or synapomorphies.

First, the accessory neural arch is consistently absent in ctenosquamates. However, as discussed above, the mosaic pattern of distribution of ANA among lower teleosts casts doubt on its homology, and therefore on the validity of its absence as a ctenosquamate character.

The second ctenosquamate feature is absence of fusion between anterior epineurals and neural arches. Fusion of epineurals is the primitive teleostean condition (e.g., *Hiodon*),

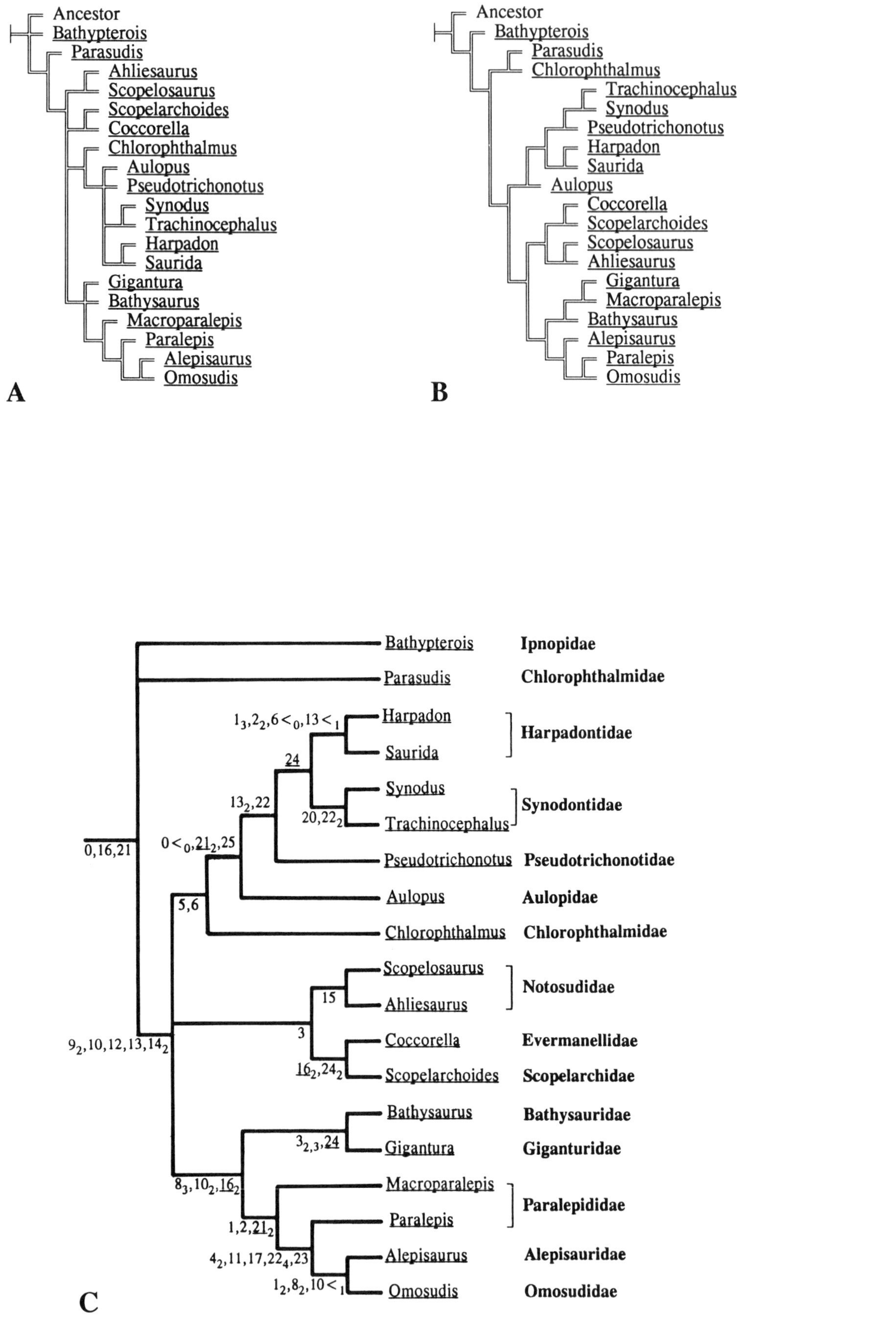
Ancestor
Bathypterois
Parasudis
Ahliesaurus
Scopelosaurus
Scopelarchoides
Coccorella
Chlorophthalmus
Aulopus
Pseudotrichonotus
Synodus
Trachinocephalus
Harpadon
Saurida
Gigantura
Bathysaurus
Macroparalepis
Paralepis
Alepisaurus
Omosudis
A
Ancestor
Bathypterois
Parasudis
Chlorophthalmus
Trachinocephalus
Synodus
Pseudotrichonotus
Harpadon
Saurida
Aulopus
Coccorella
Scopelarchoides
Scopelosaurus
Ahliesaurus
Gigantura
Macroparalepis
Bathysaurus
Alepisaurus
Paralepis
Omosudis
B
Bathypterois Ipnopidae
Parasudis Chlorophthalmidae
Harpadon
Saurida
Harpadontidae
Synodus
Trachinocephalus
Synodontidae
Pseudotrichonotus Pseudotrichonotidae
Aulopus Aulopidae
Chlorophthalmus Chlorophthalmidae
Scopelosaurus
Ahliesaurus
Notosudidae
Coccorella Evermanellidae
Scopelarchoides Scopelarchidae
Bathysaurus Bathysauridae
Gigantura Giganturidae
Macroparalepis
Paralepis
Paralepididae
Alepisaurus Alepisauridae
Omosudis Omosudidae
C

FIGURE 11 (opposite page).—Cladograms of aulopiforms based on the data in Table 6. Family names in current use for the genera in our sample are added after the generic names in C. A, Strict consensus of 24 equally parsimonious shortest trees (length 74 steps, CI 0.68, RI 0.76) found the implicit enumeration option in Hennig86 (Farris, 1988), rooted on the "ancestor" in Table 6. B, Single shortest tree found by three-item analysis with fractional weighting (Nelson and Ladiges, 1992b) using the program TAX, version 3.1 (Nelson and Ladiges, 1992a). The matrix produced by TAX contains 1434 characters; weights were multiplied by a factor of 100, and the matrix was analyzed by PAUP version 3.0 (Swofford, 1990). The PAUP tree has a length of 135351, CI 0.857, RI 0.834. When fitted to our data, the tree has a length of 82, an increase of seven steps or 9.3% over the shortest trees, but it is interesting in that it maintains a monophyletic Chlorophthalmidae (*Chlorophthalmus, Parasudis*), it places notosudids, evermanellids, and scopelarchids with alepisauroids, and it presents a novel pattern within alepisauroids. C, A consensus of two among the 24 equally parsimonious shortest trees combined in A. The two trees were selected from the 24 by discarding three sets of trees. First, we discarded trees in which *Pseudotrichonotus* is placed as the sister group either of *Aulopus* + Harpadontidae + Synodontidae or of Synodontidae alone, because both of those patterns are contradicted by other characters not considered here (Johnson et al., 1989). Second, we discarded trees in which *Gigantura* is placed as the sister group of paralepids + *Alepisaurus* + *Omosudis;* that pattern is favored only by state 1 of character 18 (some unattached epipleurals posteriorly),whereas the link between *Gigantura* and *Bathysaurus* shown here is favored by state 1 of character 24 (very few caudal vertebrae) and by possible homology between states 2 and 3 of character 3 (all or most of the epineurals originate on the centrum). Third, we discarded trees in which the group comprising *Coccorella* and *Scopelarchoides* is placed as the sister group of alepisauroids. The latter grouping is supported by state 2 of character 16 (epipleurals extend forward to V1), whereas the grouping shown here is supported by state 1 of character 3 (epineurals on ~V5-15 originate on the centrum or parapophysis); both seem to be "good" characters, unique and unreversed in their respective trees. Which of them is homoplastic might be discovered by checking further members of the families concerned, but at present our choice of the grouping shown is arbitrary, although it agrees with Nelson's result in B. Characters supporting the nodes are indicated by numbers referring to Table 6, with change from state 0 to state 1 implied unless otherwise indicated. Character states are in subscript; inferred reversals are indicated by <; and characters appearing twice on the cladogram (homoplasies) are underlined. Autapomorphies (including homoplasy and/or reversal) in terminal taxa are not indicated. Presence of ANA (state 0 of character 0) in synodontoids is resolved as a reversal, and epicentrals of *Alepisaurus* and *Omosudis* (state 1 of character 10) are resolved as secondarily reacquired.

and it persists in elopomorphs, some clupeomorphs, on a few vertebrae in *Esox,* on many in argentinoids, and on several or many in stomiiforms and some aulopiforms (Tables 3-5). It never occurs in ctenosquamates.

Myctophiforms: The pattern of the epineurals in the myctophiforms that we have sampled (Table 5) is a generalized one, and we see nothing in the intermusculars that bears on the problem of myctophiform monophyly (Johnson, 1992).

Intermusculars in Acanthomorph Teleosts

Tables 7 and 8 record the intermuscular series in a range of acanthomorphs, including *Polymixia,* with which this paper began. The overall pattern of the intermusculars in *Polymixia,* with epineural bones in a series beginning behind the head, epipleural bones in the middle part of the trunk, and a series of epicentral ligaments from V2 back into the caudal region with rod-like cartilages in them from V3-16, is not precisely matched in any other teleost that we have seen. Nevertheless, there is no problem of homology between the three series in *Polymixia* and those in lower teleosts. The only major differences in *Polymixia* concern the anterior epineurals because the bone that we interpret as the first epineural is enlarged and lies in the horizontal septum, and the epineurals of V3-10 originate on the parapophysis or centrum. In nonacanthomorphs (Tables 3-5) anterior epineurals generally originate on the neural arch and lie well above the horizontal septum. Some anterior epineurals originate on the centrum in a few nonacanthomorphs, e.g., *Xenomystus, Anchoa, Esox,* notosudids, scopelarchids (Tables 3-5), and in the aulopiform *Scopelosaurus* some anterior epineurals (on ~V6-11) originate on the parapophyses, just as in *Polymixia.* In the osteoglossoid *Arapaima,* where the first epineural is on V21 (Taverne, 1977b:229) or V22-24 (pers. obs.), the first few epineurals also originate on the parapophyses (Taverne, 1977b, fig. 141). The distal parts of the first one or more epineurals are displaced ventrally relative to their successors in argentinoids, several aulopiforms, and in *Neoscopelus* (notation b_1 or f_1 in Tables 4, 5), although they lie above the horizontal septum.

Thus, the ventrally displaced anterior epineurals of *Polymixia* are not entirely original in comparison with those of nonacanthomorphs. In any case, the intermuscular bones on V3-8 in *Polymixia* must be epineurals because there are epicentral ligaments below them. The only question of homology, therefore, concerns the intermuscular on V1: is it the serial homologue of the epineural bones on succeeding vertebrae, which it resembles in structure (bone), or of the epicentral ligaments on succeeding vertebrae, which it resembles in position (horizontal septum)? There is no decisive test to settle problems of serial homology, like the congruence test for problems of homology as synapomorphy, but ontogeny is the most useful criterion (Mabee, 1988, 1993). As Table 1 shows, the bone on V1 of *Polymixia* develops in series with the epineurals and is fully formed at 12 mm SL, before the epicentrals are recognizable. Furthermore, as Tables 3-5 show, there is no lower teleost in which the epicentral series contains just one bone on V1 followed by a series of ligaments; when epicentral bones occur, they form a complete series anteriorly (*Megalops,* clupeomorphs, gonorynchiforms, *Parasudis*). Both lines of evidence indicate that the bone on V1 of *Polymixia* is an epineural displaced ventrally into the horizontal septum, not an epicentral. We take that conclusion as the starting point for our comparisons of the intermusculars in *Polymixia* with those of other acanthomorphs, and we use a description of the intermusculars in holocentrid beryciforms and centropomid perciforms as an introduction to those comparisons.

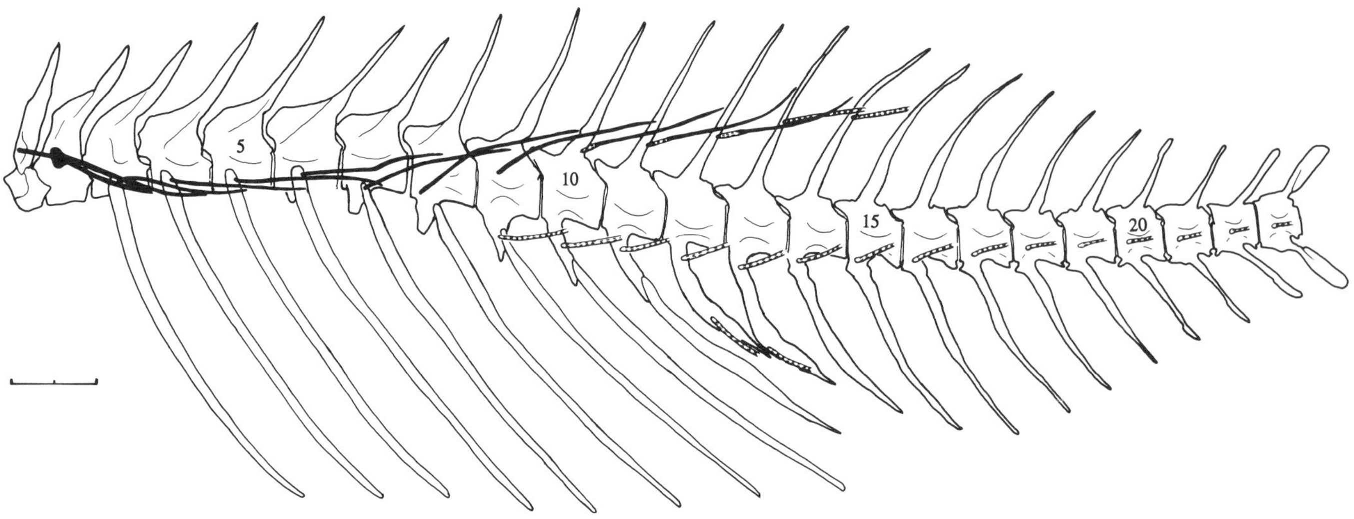

FIGURE 12.—Intermuscular system of bones and ligaments in *Holocentrus diadema* Lacépède, based on BMNH 1960.3.15.117 (40 mm SL). The intermuscular bones are in solid black, the ligaments are shaded. Every fifth vertebra is numbered. (Scale bar in mm.)

HOLOCENTRIDS

As stated in the "Introduction," all Recent acanthomorphs except *Polymixia* have only one series of ossified intermusculars, or no such bones. Holocentrids (Figures 12, 13) are used here to exemplify the primitive form of that acanthomorph condition. Within holocentrids, we construe *Holocentrus* in a broad sense, as including species currently placed in *Sargocentron* or *Adioryx,* as Stewart (1984) was unable to find any apomorphies distinguishing those species, and they are evidently paraphyletic relative to the two species (*H. ascensionis, H. rufus*) currently grouped in *Holocentrus* by an apomorphic ear/swimbladder linkage. The description that follows is based on the illustrated counterstained specimen of *Holocentrus diadema* Lacépède, a counterstained specimen of *H. spiniferum* Forskål (BMNH 1960.3.15.169), three alizarin-stained specimens of *H. vexillaris* (USNM 269533), and counterstained and alizarin-stained specimens of *Ostichthys* and *Myripristis*. In the *Holocentrus* illustrated (Figures 12, 13), the series of ossified intermusculars extends over 12 vertebrae. On V1 (cf. Rosen, 1985, fig. 20, where it is called an epineural) the bone originates rather high on the (autogenous) neural arch, extends posterolaterally in the horizontal septum, and ends immediately below the lateral line nerve at the level of the junction between the third and fourth centra. This bone has the same relations as the epineural on V1 in *Polymixia,* but it differs from that in one respect: its distal tip is surrounded by (but is not continuous with) an irregular superficial plate of cartilage (Figure 13). The intermuscular on V2 in *Holocentrus* has an enlarged head (Rosen, 1985, fig. 20, where it is called a pleural rib). It originates in a socket at the extreme base of the (fused) neural arch, extends posterolaterally in the horizontal septum, and ends beneath the lateral line nerve. Like the first intermuscular, the distal tip of the second is surrounded by an irregular but smaller superficial plate of cartilage (Figure 13). The next five intermusculars, on V3–7, each originate on the head of the rib, and their distal end is successively further above

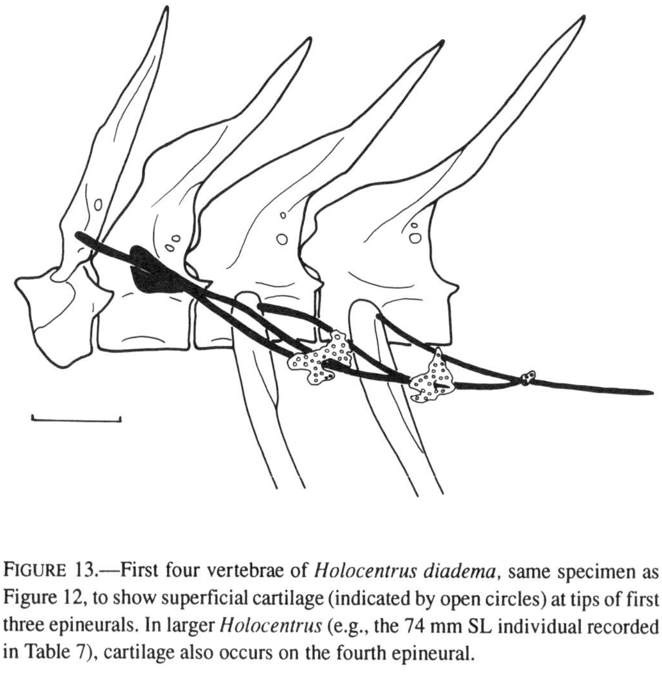

FIGURE 13.—First four vertebrae of *Holocentrus diadema,* same specimen as Figure 12, to show superficial cartilage (indicated by open circles) at tips of first three epineurals. In larger *Holocentrus* (e.g., the 74 mm SL individual recorded in Table 7), cartilage also occurs on the fourth epineural.

the lateral line nerve; the distal tip of the third is associated with a small, superficial blob of cartilage (Figure 13), but there is no cartilage at the tip of the others. On V8–10 the intermusculars originate at successively higher points on the parapophysis and centrum rather than on the rib (Figure 12): at the base of the parapophysis on V8, on the centrum, by means of a short ligament, on V9, and on the base of the neural arch, by means of a ligament, on V10. The distal ends of these intermusculars lie at successively higher levels above the horizontal septum, a condition first noted in holocentrids by Owen (1846:66; also Günther, 1859:24, 38, on *Myripristis* and *Holocentrus*). The last two intermuscular bones are attached by long ligaments to neural spines of V11–12. Like their predecessors, they end distally at successively higher levels, with the last one terminating closer to the tip of the neural spine than to the lateral line. On V13 and V14 there are epineural ligaments on the neural spines; the epineural ligament on V13 has a sliver of included bone distally on the right side only.

In holocentrids, as in *Polymixia* (Figure 1, Plate 2A–D) and most lower teleosts, there are epicentral ligaments extending from the vertebrae in the horizontal septum (Plate 2E,F). The first is on V9, and it originates on the parapophysis, just in front of the rib head. The last two rib-bearing vertebrae, V10–11, have closed haemal canals, and the epicentral ligament originates above the rib on the haemal arch. On the first three caudal vertebrae, V12–14, the epicentral ligament originates at successively higher points on the base of the haemal spine, and on V15–17 it is on the anteroventral part of the centrum. On more posterior vertebrae, the point of origin of the ligament shifts toward the middle of the centrum, and the last detectable epicentral is on V24 (PU5).

Posterior oblique ligaments (POTs) are present in *Holocentrus*. The most anterior one originates anteroventral to the epicentral ligament on the haemal spine of the second caudal vertebra (V13), passes forward beneath three epicentral ligaments, and inserts on the lateralis musculature together with the epicentral ligament of V9 (the most anterior member of the epicentral series). On succeeding caudal vertebrae the POTs have the same pattern, inserting with the epicentral of the vertebra four in front. On vertebrae behind V13, the point of origin of the POT moves up the haemal arch on to the anteroventral corner of the centrum, and more posteriorly (at about V17) the origin of the POT shifts across the intervertebral joint on to the posteroventral corner of the preceding centrum. The last detectable POT is on V25 (PU4). The POTs in holocentrids are dorsoventrally flattened, strap-like ligaments, about half the size of the epicentral ligaments anteriorly, and about equal in size to them posteriorly.

There are generally no epipleural bones in holocentrids, but in specimens of *H. diadema*, *H. spinifer*, and *H. vexillaris* there are ligaments in the epipleural position on haemal spines of the first two caudal centra (V12–13), and in one individual (the largest, 77 mm SL) among the three in USNM 269533 (*H. vexillaris*) there is a sizeable bone in the first of these ligaments on both sides of the specimen. We have not seen an epipleural bone in any other holocentrid or beryciform, although we have examined many cleared-and-stained specimens as well as dried skeletons and radiographs of large specimens, nor is one recorded in previous descriptions (Starks, 1904a; Zehren, 1979; Stewart, 1984), and we therefore regard the single epipleural bone in this individual as anomalous or atavistic.

Other holocentrids show only trivial differences from the pattern described above and shown in Figures 12 and 13. A counterstained specimen of *H. spiniferum* (76 mm SL) has 13 rather than 12 intermusculars, with that on the 8th vertebra attaching to the rib rather than the parapophysis, and that on the 11th vertebra attaching directly to the neural spine rather than by ligament, and it has the fourth intermuscular ending in a small superficial plate of cartilage, like the first three in the illustrated specimen (Figure 13).

A counterstained specimen of *Myripristis pralinius* (BMNH 1974.5.25.699, 65 mm SL) and the alizarin-stained *M. murdjan* illustrated in Plate 2E,F indicate that intermuscular bones and ligaments of *Myripristis* show the same pattern as in *Holocentrus* in every detail except that no cartilage is detectable at the tips of anterior intermusculars. As in *Holocentrus*, epipleural ligaments are present on the first two caudal haemal spines.

CENTROPOMIDS

In Table 8 we record the intermusculars in two species of the centropomid *Lates*, one *L.* sp. (from a cleared-and-stained lot of 11 planktonic young of *L. longispinis* or *L. niloticus* from near Kigoma, Lake Tanganyika; introduced populations, determination by P.H. Greenwood) and an adult *L. calcarifer*. *Lates* sp. is illustrated in Figure 14. In our specimens there are four (*L. calcarifer*) or five (*L.* sp.) intermuscular bones; Greenwood (1976) found eight intermusculars in adult *L. niloticus* and five in *L.* (*Luciolates*) *stappersi*. These bones have the same distribution as anterior epineurals in holocentrids: all lie in the horizontal septum, with the first two originating on the neural arch (epineurals of Greenwood, 1976) and the remainder on the rib (epipleurals of Greenwood, 1976). Behind these intermuscular bones there is a gap of four (*L.* sp.) to seven (*L. calcarifer*) vertebrae before the first epicentral ligament, which originates on the parapophysis (forming a closed haemal arch) of V10 in *L.* sp., and on that of V12, the first caudal vertebra, in *L. calcarifer*. In *L.* sp. there are nine more epicentral ligaments and in *L. calcarifer* there are seven more epicentral ligaments, with the last on V19 in both species. Passing posteriorly along the column, the origin of the epicentrals moves from the base of the haemal spine to the middle of the centrum, as shown in Figure 14.

Posterior oblique ligaments (POTs) are well developed in *Lates*. In our juvenile *L.* sp., the first POT originates on the parapophysis of V7 and inserts on the distal part of the intermuscular of V4. The POT from the parapophysis of V8

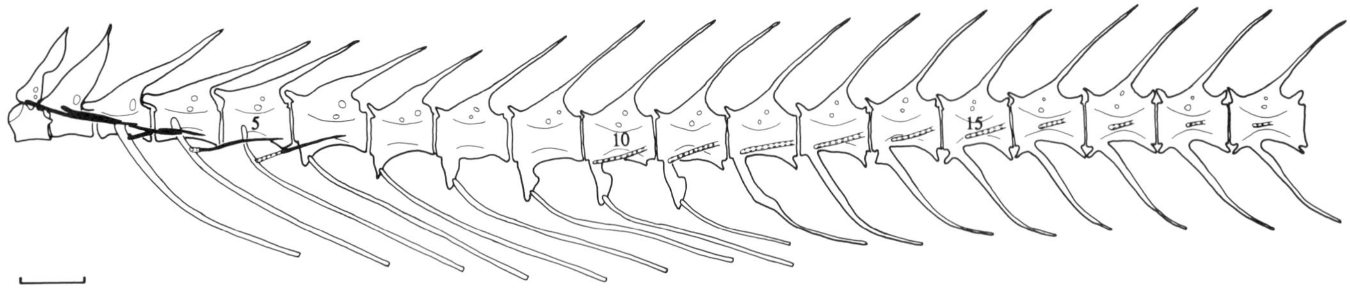

FIGURE 14. Intermuscular system of bones and ligaments in *Lates* sp. (*L. longispinus* Worthington or *L. niloticus* (L.)), based on unregistered BMNH pelagic larvae, 35–40 mm SL. The intermuscular bones are in solid black, the ligaments are shaded. Every fifth vertebra is numbered. (Scale bar 1 mm.)

inserts on the tip of the intermuscular of V5 (the last intermuscular, which is attached to the centrum by ligament), and POTs from V9–12 insert at the lateral junction of the myoseptum and horizontal septum. The point of origin of POTs shifts from the parapophysis (placed anteriorly on the vertebra) to the posterior part of the preceding centrum at about V11, so that in the anterior caudal region there is an epicentral ligament originating anteriorly on each centrum and a POT originating posteriorly, the pattern described by Kafuku (1950). The first epicentral ligament, on V10, inserts with the POT from the centrum of V11. The last obvious POT originates on PU7, with the last epicentral ligament, and inserts with the epicentral ligament from PU8. In our adult *Lates calcarifer*, the first detectable POT originates on V9.

There are no epipleural bones or ligaments in *Lates*.

Table 8 also records the intermusculars in the centropomid *Psammoperca waigiensis*. The pattern in *Psammoperca* agrees with that in *Lates* in most details. There are eight epineurals, as in adult *L. niloticus*, but epicentrals begin on V10 (the last abdominal), on the same vertebra as the first epicentral in *Lates* sp., so that there is a gap of only one vertebra between the epineural and epicentral series. The first POT originates on the parapophysis of V6 and inserts on the fourth epineural. As in *Lates*, there are no epipleurals in *Psammoperca*.

Centropomus (we checked *C. ensiferus*) has six (our *C. ensiferus*) or seven (Fraser, 1968) epineural bones and the first epicentral ligament originates on V10 (the last abdominal). The first POT also originates on V10, and inserts on the epineural of V6.

COMPARISON BETWEEN THE INTERMUSCULAR BONES AND LIGAMENTS IN *Polymixia*, HOLOCENTRIDS, AND CENTROPOMIDS

Figure 15A–C summarizes the condition of the intermuscular bones and ligaments in *Polymixia*, a holocentrid, and a centropomid. As shown above, the pattern in *Polymixia* is readily homologized with that in nonacanthomorph teleosts (aulopiforms in particular), and the only unique difference is that the first epineural of *Polymixia* is displaced ventrally into the horizontal septum.

The intermusculars of holocentrids differ from those of *Polymixia* in six ways: (1) the first five to seven epineurals (rather than just the first) lie in the horizontal septum; (2) epineurals on V3 to V7–8 originate on the rib, not the parapophysis; (3) epicentral ligaments are absent on V1–8 (rather than just V1); (4) there are no cartilage rods in the epicentral ligaments, although there may be superficial cartilage plates at the tips of the first few epineurals; (5) the first POT is on the second caudal vertebra (V13) and inserts on the first epicentral ligament, on V9 (in *Polymixia* the first POT is on V9 and inserts on the epicentral ligament of V5); and (6) there are no epipleural bones in holocentrids (except a single bone in one anomalous individual), just epipleural ligaments on the first two caudal vertebrae. We interpret five of these six differences as derived conditions in holocentrids relative to those in *Polymixia*. In holocentrids the second to seventh epineurals have descended into the horizontal septum (1), so eliminating epicentrals of the anterior vertebrae (3) and anterior POTs (5); origin of the epineurals on V3 to V7 or V8 has shifted from parapophysis to rib (2); and the epipleural series is greatly reduced (6). The other difference (4), concerning cartilage in the epicentrals or at the tip of the anterior epineurals, may merely reflect autapomorphic conditions in both *Polymixia* and holocentrids; a wider survey of double-stained acanthomorphs, particularly of relatively large specimens, is necessary before anything useful can be concluded.

The intermusculars of centropomids differ from those of

FIGURE 15 (opposite page).—Comparative summary of intermuscular patterns in acanthomorphs. A, Primitive euacanthomorph pattern in *Polymixia lowei*, as in Figure 1; B, Primitive acanthopterygian pattern in *Holocentrus diadema*, as in Figure 12; C, *Lates* sp., as in Figure 14, exemplifying the pattern we take to be primitive for percomorphs; D, *Pseudanthias squamipinnis* (Peters), based on BMNH 1975.4.5.14 (29 mm SL), exemplifying the pattern we take to be derived in percomorphs. Ossified intermusculars are in solid black, cartilaginous epicentrals in *Polymixia* are stippled. (Scale bars in mm.)

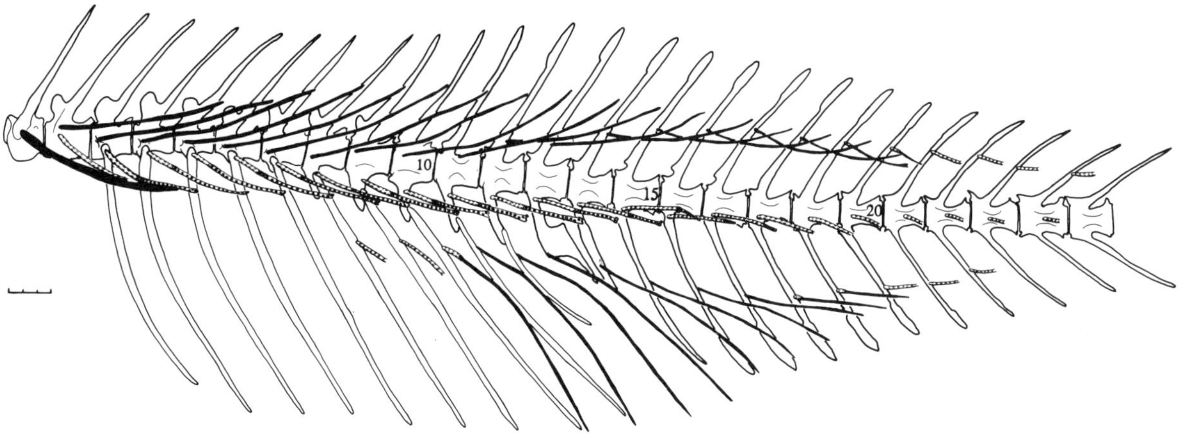
10
15
20

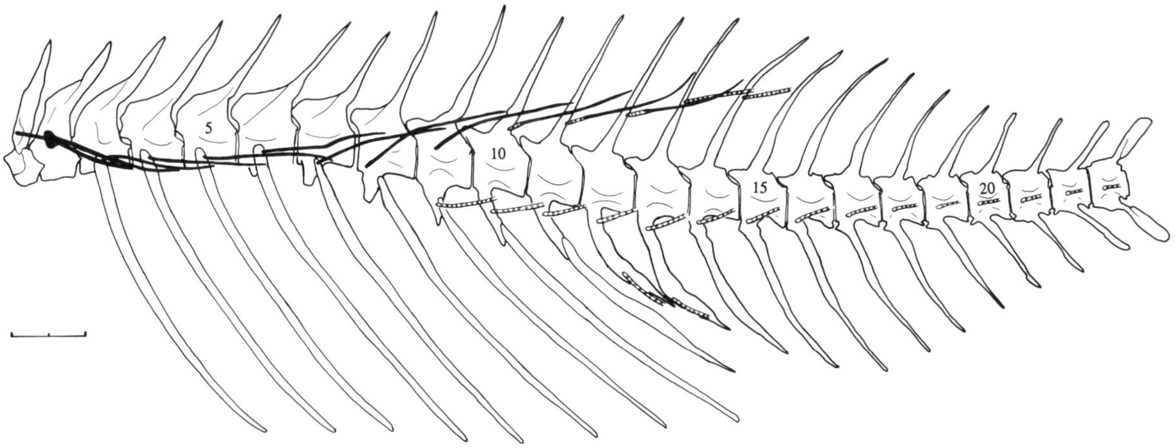
5
10
15
20

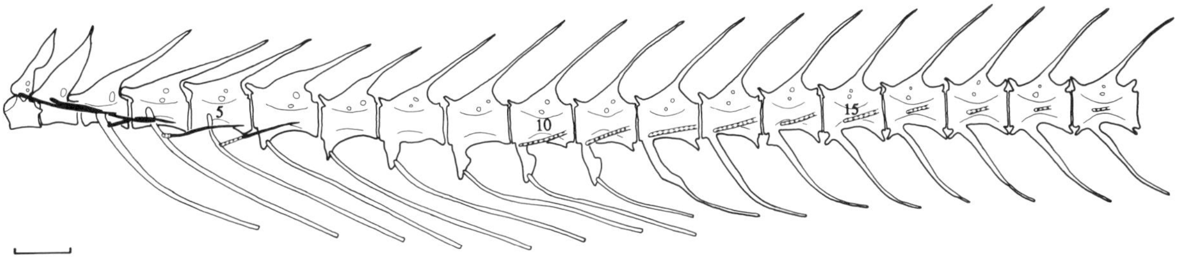
5
10
15

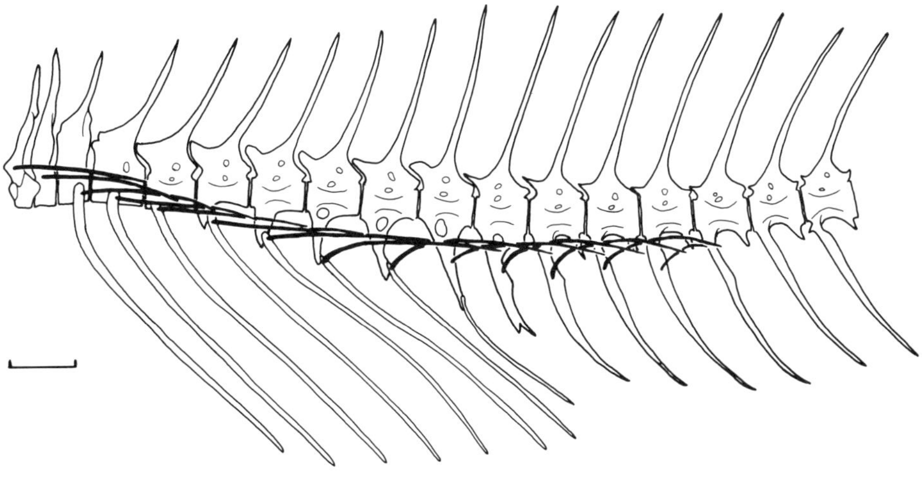

holocentrids in five ways: (1) epineurals are confined to the first four (*Lates calcarifer*) to eight (*L. stappersi, Psammoperca*) vertebrae, and they all lie in the horizontal septum; (2) there is no cartilage at the tip of anterior epineurals; (3) epicentral ligaments are absent on V1–9 (*Psammoperca, Centropomus, Lates* sp.) or V1–11 (*Lates calcarifer*) so that there is a gap of from one (*Psammoperca*) to seven (*L. calcarifer*) vertebrae between the last epineural and the first epicentral; (4) the first POT originates on V6 (*Psammoperca*), V7 (*L.* sp.), V9 (*L. calcarifer*), or V10 (*Centropomus*), in front of or beneath (*Centropomus*) the first epicentral ligament, and the most anterior POTs insert on epineural bones rather than epicentral ligaments; and (5) there are no epipleural ligaments. We interpret four of these five differences as derived conditions in centropomids relative to those in holocentrids. In centropomids the epineural series is truncated caudally (1) and the epicentral series is further truncated rostrally (3) so that the two series no longer overlap; the POTs have extended rostrally beyond the epicentrals and have established a new relationship with the epineurals in the horizontal septum (4); and the epipleural series is lost (5). The remaining difference, concerning absence of cartilage at the tips of anterior epineurals (2), probably reflects autapomorphic conditions in holocentrids, as surmised above.

THE INTERMUSCULAR SERIES AND ASSOCIATED STRUCTURES IN OTHER ACANTHOMORPH TELEOSTS

Tables 7 and 8 record the distribution and form of the intermuscular bones and ligaments in a range of acanthomorphs. In the corresponding section on nonacanthomorphs (p. 11) we commented separately on the three series of intermusculars in those fishes. Acanthomorphs lack epipleurals (excepting *Polymixia*, ligaments on the first two caudal vertebrae in holocentrids, and questionable ligaments in the lampridiform *Velifer*, p. 43) so there is no need for further comment on those. We discuss the two other series, the epineurals and epicentrals, together, because we believe that they are indistinguishable in some acanthomorphs. We also comment on acanthomorph POTs, ribs, and certain other features recorded in the tables.

EPINEURALS AND EPICENTRALS.—The preceding comparisons between the intermusculars of *Polymixia*, holocentrids, and centropomids showed three patterns of the epineurals and epicentrals (Figure 15A–C). We regard these three conditions as a morphocline. As Tables 7 and 8 show, the pattern in *Polymixia* is approximated only in lampridiforms among acanthomorphs. Lampridiforms are discussed separately below, and the *Polymixia* pattern needs no further comment here.

The holocentrid pattern is matched only in other beryciforms, notably in the berycid *Centroberyx* and the anomalopid *Anomalops* (Table 7). Essentials of this pattern are a series of epineurals that originate on ribs anteriorly and ascend to neural arches and spines posteriorly, and a series of epicentral ligaments that begin on the posterior abdominal vertebrae, at about V9. Because the epicentral series overlaps the epineural series over several vertebrae there is no problem of serial homology. Variants of the holocentrid pattern recorded in Table 7 include that in the trachichthyid *Trachichthys* and the anomalopid *Photoblepharon*, where the epineural series is interrupted, with bones on the first two vertebrae and ligaments beginning at V6–7, and that in *Monocentris*, where epineurals do not begin until V8. Stephanoberyciforms (Table 7) show a different pattern, with few epineurals (the eight in *Gibberichthys* and 10 in *Rondeletia* are exceptional) and no epicentrals (except for a questionable bone on the centrum of V2 in our specimen of *Barbourisia*, probably an individual anomaly).

The pattern in centropomids, with a gap of one to several vertebrae between the last epineural and the first epicentral, is matched in three other percoids in Table 8: *Morone* (Moronidae), *Doederleinia* (Acropomatidae), and *Drepane* (Drepanidae), although in *Drepane* epineurals extend to the second caudal vertebra (where they originate on the haemal arch) and epicentrals begin on the fifth caudal vertebra. The centropomid pattern also is matched in the zeiform *Xenolepidichthys* (Table 7), in the mugiloids *Agonostomus* (Table 8) and *Liza* (Table 8), and in a 14 mm SL larva of the scombroid (sensu Johnson, 1986) *Sphyraena* (Table 8).

A much more common pattern among the percomorphs (sensu Johnson and Patterson, 1993) in Table 8 is that shown, for example, by the atherinomorphs *Bedotia, Menidia,* and *Exocoetus,* by the serranid *Pseudanthias* (Figure 15D), the caproids *Antigonia* and *Capros,* the acanthuroid *Siganus* (and other acanthuroids illustrated in Tyler et al., 1989, figs. 23–29), the anabantoid *Colisa,* the cottoid *Cottus,* and the carangoids *Caranx* and *Rachycentron.* In this pattern there is only a single, uninterrupted series of intermusculars, originating on the ribs on anterior rib-bearing vertebrae (except in *Menidia* and *Exocoetus,* where they are on the parapophyses), on the parapophysis and/or haemal arch more posteriorly, and usually ascending to the centrum on caudal vertebrae. The series may be all in bone (e.g., *Elassoma, Bedotia, Exocoetus, Pseudanthias, Cottus*) or may be in bone anteriorly and ligament caudally (e.g., *Menidia,* caproids, *Siganus, Colisa,* carangoids). A similar pattern occurs among nonpercomorph acanthomorphs (Table 7) in gadiforms and in the stephanoberyciform *Rondeletia.* There is an obvious problem of homology with the elements (bones, ligaments) of this pattern: are they all epineurals, serial homologues, or are anterior members of the series epineurals and posterior members epicentrals (the ligaments, perhaps), in which case they are not all serial homologues?

We have come to the conclusion that in most cases this question is meaningless, and we have decided to call them all epineurals (they are entered as such in Tables 7 and 8). The question is analogous to the problem of "phylogenetic fusion" between bones (discussed by Patterson, 1977), with the added dimension or complication of serial homology. Where there is evidence of ontogenetic fusion between two bones, there is no real difficulty. A compound name for the resultant single bone sometimes may be thought appropriate (e.g., the intertemporo-supratemporal of *Polyodon*, Grande and Bemis, 1991:6), or the compound name is not found necessary, as with the teleostean pterotic (ontogenetic fusion between dermopterotic and autopterotic in some teleosts, Patterson, 1977:98), palatine (ontogenetic fusion between dermopalatine and autopalatine in some teleosts, Patterson, 1977:100), and quadrate (ontogenetic fusion between quadrate and quadratojugal in some teleosts, Patterson, 1977:100). But when there is no evidence of ontogenetic fusion, and a single bone in one form or group is held to be homologous with two bones in outgroup taxa, is it better to assume that the two bones have "fused" in phylogeny, and so prefer the compound name, or has one of the two bones been lost or "captured" by the other, so that a single name is more appropriate? Ørvig (1962) and Nelson (1969, 1973) concluded that distinction between phylogenetic fusion and loss in such cases was meaningless ("merely semantic," Nelson, 1973:339) and opted for fusion "for want of a better word" (Nelson, 1969:10). Patterson (1977:96) argued that the hypothesis of phylogenetic fusion was preferable, "when supported by evidence of ontogenetic fusion in some forms," because the alternative, i.e., loss of one bone through "capture" or "invasion" by the other, was irrefutable and therefore untestable.

In our case, the question of distinction between epineurals and epicentrals concerns serial homology as well as taxic homology and is, therefore, different from discriminating between fusion or loss for two bones. As with fusion between bones, ontogeny provides the best test, and fusion and loss hypotheses are both testable because each makes different predictions. For example, we have evidence of ontogenetic fusion between the epineural and epicentral series in the primitive scombroid (sensu Johnson, 1986) *Sphyraena* (Table 8). In a 14 mm SL larva, there are only three epineural bones, on V1–3, and the epicentral series of ligaments occupies V9–15. In a 79 mm SL subadult, the epineural series of bones occupies V1–7 (that on V7 is attached by ligament) and that series is continued caudally by ligaments on V8–19. Thus, in *Sphyraena* we have evidence that the epineural series has developed rostrocaudally and has become continuous with a pre-existing series of epicentral ligaments. In Table 8, we enter bones on V1–7 of the subadult as epineurals and ligaments on V8–19 as epicentrals. But it will be evident that there is some arbitrariness here, i.e., is the ligament on V8 (which does not exist in our 14 mm larva) "really" the last epineural or is it the first epicentral? We have no solution. And in cases like *Menidia*, caproids, and carangoids, where in our specimens there is continuity between intermuscular bones anteriorly and ligaments posteriorly and where we have no ontogenetic series, no rational decision is yet possible on whether some or all of the ligaments are "truly" epicentrals. We expect that ontogeny will be decisive here. For example, it will be interesting to learn from ontogeny whether the pattern in gadiforms like *Raniceps* and *Gaidropsarus* (Table 7) is the result of craniocaudal development of a single series of structures (epineurals) or is the result of coalescence of two separate series (epineurals and epicentrals, as in *Sphyraena*). Because we have not found epicentral ligaments in any paracanthopterygian, we predict that larvae of these gadiforms will show craniocaudal development of a single series of epineurals.

There are two instances among the percomorphs in Table 8 where we have entered overlapping series of epineurals and epicentrals: the scombrid *Scomberomorus* and the polynemid *Polydactylus*. Westneat et al. (1993) have described intermuscular bones and ligaments in *Scomberomorus* and several other scombrids. We confirm their description (p. 188) of the condition in *Scomberomorus:* "An AOT [our epicentral ligaments] is a single band of fibers attaching to the posterior edge of the epipleural ribs [our epineurals] in the anterior region of the body. The anterior epipleural ribs have reduced AOTs in comparison to the thicker AOTs that attach to the posterior epipleural ribs." But Westneat et al. (1993:193) note that in *Sarda, Euthynnus,* and *Thunnus* "most AOTs ... surround epipleural ribs. ... Bands of AOT fibers occur on both the posterior and anterior edge of each epipleural rib." Thus, in these scombrids the epineural bones appear to be ossifications within or surrounded by ligaments, rather than distinct from them (as in *Scomberomorus*). As with the ligament on V8 in *Sphyraena* (above), it will be evident that the distinction between epineurals and epicentrals in scombrids is more-or-less arbitrary. Westneat et al. (1993:188) suggest that "epipleural ribs are likely to be ossified serial homologues of the AOTs that occur in the main horizontal septum." This is an accurate statement of the topographic relations between intermuscular bones and ligaments in many percomorphs, but we hope we have shown that in the phylogenetic sense the single series of intermusculars in those fishes comprises two series, the epineurals and epicentrals, that are distinct in polymixiids, beryciforms, centropomids, and some other percomorphs. Ontogenetic evidence is necessary to determine whether the two series "fuse" during development (as in *Sphyraena*) or are no longer distinguishable.

The second instance of overlapping epineural and epicentral series in Table 8 is the polynemid *Polydactylus*. Here there is a series of decended epineural bones on V1–7, and that series is continued caudally by a series of ligaments on V8–15. The point of origin of the last two epineural bones ascends from rib to parapophysis, a frequent pattern for posterior epineurals in percomorphs (e.g., *Psammoperca, Pseudanthias;* Figure 15D, Table 8). Following these bones is the series of ligaments on V8–15, but the anterior members of that series (V8–10) originate not on the parapophysis but more ventrally, on the rib, and the remainder originate on the haemal spine. These ligaments do not lie in the main horizontal septum, beneath the lateral line, but in the ventral secondary horizontal septum (described by Westneat et al., 1993, in scombrids), which is particularly well developed in *Polydactylus*. Furthermore, the first two of these ligaments, originating on the ribs of V8–9, are weaker and more diffuse than their successors, implying that the series of ligaments expresses a different developmental gradient from the epineural series of bones. Above these ligaments, there is another series of ligaments beginning on V10 and entered as epicentrals in Table 8. The first of these originates at the anteroventral corner of the centrum of V10, and the points of origin of its successors ascend to the centrum in the normal pattern of epicentral ligaments; the series extends into the caudal region, to PU4. *Polydactylus* also has a series of "neoneural" ligaments (notation Nn in Table 8; see below).

The situation in *Polydactylus* does not match any other genus in Table 8, and we do not yet fully understand it. The interpretation that we favor is that the epineurals and epicentrals show the pattern that we take to be primitive for percomorphs, as described above in centropomids, *Morone*, mugilids, and larval *Sphyraena*, where the epineural and epicentral series are discontinuous; in *Polydactylus* the discontinuity would be the gap between the last epineural bone on V7 and the first epicentral ligament on V10. Ligaments on the ribs of V8–10 and the haemal spines of V11–15 lie in the ventral secondary horizontal septum, in the position of epipleurals, and the fact that anterior members of the series are weaker than their successors implies that they develop in the same way as epipleurals, rostrally and caudally from the region of the first caudal vertebra. If they are epipleural ligaments, we do not believe that they are homologous with the primitive epipleural series, which is absent in all percomorphs and is represented in acanthopterygians only by two ligaments in holocentrids. We therefore suggest that these ligaments in *Polydactylus* are a neoformation, analogous to the "neoneural" ligaments found above the vertebral column in *Polydactylus, Ammodytes*, and the mullid *Upeneus* (Table 8 and below); they might be called "neopleurals." Further work is necessary to determine whether a similar series of ligaments occurs in other percomorphs and whether the interpretation just proposed is valid.

Owen (1866:44), in first discriminating the three series of intermuscular bones, wrote that "in *Glyphysodon* [= *Abudefduf*, Pomacentridae] the epipleurals [= epineurals] are anchylosed to the ribs." The pomacentrid *Dascyllus* is included in Table 8 and confirms Owen's observation. An epineural is fused to every rib. In a 12 mm SL specimen of *Abudefduf biocellatus*, all the epineurals are ossified, but none is yet fused with a rib.

POSTERIOR OBLIQUE LIGAMENTS (POTs).—Our survey of POTs in acanthomorphs has been more cursory than our survey of their epineurals and epipleurals.

POTs occur in *Polymixia* and lampridiforms, and those two taxa are unique among acanthomorphs in having POTs that originate on abdominal vertebrae and insert on anterior epicentral ligaments (the first POT originates on V9 in *Polymixia* and *Velifer* and inserts on the epicentral of V5 in *Polymixia* and of V3 in *Velifer*).

We have found no POTs in any paracanthopterygian in our sample, which included percopsiforms, ophidiiforms, gadiforms, batrachoidiforms, and lophiiforms. In stephanoberyciforms, POTs occur only in melamphaids, where the first originates on V9 in *Poromitra*, on V10 in *Scopelogadus*, and on V12 in *Melamphaes*. Because these fishes have no epicentrals (like all stephanoberyciforms), POTs of melamphaids do not cross any other ligaments in passing anterolaterally to the lateralis musculature.

Among beryciforms, holocentrids have POTs (described above) from V13 (second caudal vertebra) back to about PU4. The POT originating on V13 inserts on the first epicentral ligament, from V9. In *Anomalops*, the first POT originates on V11, posteroventral to the first epicentral ligament, which also originates on V11. Because there are no epicentrals anterior to this, POTs of V11–13 lie free in the horizontal septum (like those of melamphaids) as they pass forward after crossing the epicentral of V11; more posterior POTs insert on the epicentral from the vertebra four ahead of their origin.

In zeiforms, POTs show a pattern like that in percomorphs, with some anterior POTs originating on abdominal vertebrae and inserting on epineural bones. In *Parazen*, the first POT originates on V12 but extends forward past eight vertebrae to insert on the third epineural, which is unattached proximally (Table 7) but is aligned with V3. In *Xenolepidichthys*, the first POT originates on V6 and inserts on the first epineural, which is on V2.

In percomorphs (sensu Johnson and Patterson, 1993), the pattern described above in centropomids (and just mentioned in zeiforms) is common; in it some anterior POTs originate on abdominal vertebrae and insert on epineural bones. A few examples of that pattern follow. In *Serranus cabrilla* the first POT originates on V10 and inserts on the epineural of V6; in *Morone* the first POT runs from V9 to the epineural of V6; in *Sphyraena* the first POT runs from V8 to the epineural of V7 (the last epineural bone); in *Drepane* the first POT runs from V8 to the epineural of V5; in *Antigonia* the first POT runs from V7 to the epineural of V4; in *Mulloidichthys* (Mullidae) the first POT runs from V6 to the epineural of V3; in *Caranx* the first POT runs from V5 to the epineural of V3; and in *Siganus* the first POT runs from V3 to the epineural of V1.

All percomorphs named in the preceding paragraph have a series of ligaments in the caudal region (the AOTs of Kafuku, 1950, and Westneat et al., 1993), and that series of ligaments is either continuous with epineural bones anteriorly (adult *Sphyraena, Antigonia, Mulloidichthys, Serranus, Siganus*) or is separated from epineurals by a gap of several vertebrae (*Morone, Drepane*). There is another common percomorph pattern in which there are no detectable AOTs, only epineural bones, so that the anterior POTs insert on epineural bones and the posterior POTs insert on the lateralis musculature, without crossing any AOTs. Examples of that pattern include *Epigonus*

(Epigonidae), where the first POT runs from V7 to the fourth (of six) epineurals, *Cookeolus* (Priacanthidae), with the first POT from V9 to the fifth epineural, *Elassoma* (Elassomatidae), with the first POT from V12 to the epineural of V9, and *Cottus*, with the first POT from V10 to the epineural of V5.

Finally, there is a third set of percomorphs in which we have found no POTs in the cleared-and-stained specimens that we have checked. These include the percoid *Pseudanthias* (Serranidae); the polyenemoid *Polydactylus* (Polynemidae); the scorpaeniforms *Platycephalus* (Platycephalidae), *Hoplichthys* (Hoplichthyidae), *Anoplopoma* (Anoplopomatidae), and *Hexagrammus* (Hexagrammidae); the trachinoids *Champsodon* (Champsodontidae), *Trichodon* (Trichodontidae), *Trachinus* (Trachinidae), *Uranoscopus* (Uranoscopidae), *Bembrops* and *Hemerocoetes* (both Percophididae), and *Parapercis* (Pinguipedidae); the blennioids *Forsterygion* (Tripterygiidae) and *Dactyloscopus* (Dactyloscopidae); the ammodytoid *Ammodytes* (Ammodytidae); and the callionymoid *Synchiropus* (Callionymidae). A wider and more thorough survey that includes larger specimens is necessary to determine whether absence of POTs in all or any of these genera is real or artefactual, and if real, whether it has any significance in characterizing groups.

MYORHABDOI.—The only acanthomorph in Tables 7 and 8 in which we observed myorhabdoi (series of unattached intermuscular bones in the uppermost and lowermost forward flexures of the myoseptum) is the pleuronectiform *Bothus*. Bothid pleuronectiforms have five series of intermuscular bones (Amaoka, 1969, figs. 120–122, 125), which are in a pattern resembling that in *Notopterus* amongst osteoglossomorphs (Taverne, 1978, and above) and in some clupeoids (Phillips, 1942). Samarid pleuronectiforms have three series of intermusculars (Hensley and Ahlstrom, 1984; Table 8), agreeing in position with the middle three of the five series in bothids. Amaoka (1969) took the numerous intermusculars of bothids to be primitive and to be evidence that pleuronectiforms are not related to percoids. Hensley (1977) and Hensley and Ahlstrom (1984) have argued, correctly in our view, that the extra intermuscular series in bothids (and samarids) are apomorphic. *Psettodes*, the sister group of other pleuronectiforms (Chapleau, 1993), has epineurals on V1–7 in a typical percomorph pattern (B_1, B_1, B_4, B_4, B_4, B_3, B_3 in the notation of Tables 7, 8; Amaoka, 1969, fig. 118). Citharids have epineurals on V1–10 or V1–11 and paralichthyids have them on V2–10 or V2–11, attaching to the centrum anteriorly and parapophyses or haemal arches posteriorly (Amaoka, 1969).

Myorhabdoi also are present in the gasterosteiform family Aulostomidae. They were described briefly in the single Recent genus *Aulostomus* by Jungersen (1910), and they are shown to have existed in the group since at least the Eocene by their presence in the genera *Eoaulostomus*, *Jungersenichthys*, and *Synhypuralis*, from Monte Bolca, Verona, Italy. Myorhabdoi also occur in gempylids (pers. obs.).

"NEONEURAL" LIGAMENTS AND BONES.—In six percomorphs in Table 8 we record a series of bones or ligaments against the symbol "Nn" meaning "neoneural." The taxa are *Ammodytes* (Ammodytoidei), *Dascyllus* (Pomacentridae), *Polydactylus* (Polynemoidei), *Upeneus* (Mullidae), and the pleuronectiforms *Samaris* and *Bothus*. Neoneurals are all in ligament in the first four genera and are all in bone in *Samaris* and *Bothus*. In *Upeneus*, ligaments originate on the neural arch of V2–6 and rise to the neural spine on V8–14. In *Dascyllus*, the ligaments originate on the neural arch of V2–13 and rise on to the neural spine on V14–18. In *Polydactylus*, ligaments originate on the neural arch of V1–3, on the centrum of V4–8, and ascend to the neural spine posteriorly (V12). In *Ammodytes*, where the ligaments are very robust and obvious, they originate on the centra of V5–33 and ascend to the neural arch and spine posteriorly. We interpret these series of ligaments as neoformations, nonhomologous with anything in lower acanthomorphs and nonacanthomorphs. We are satisfied that they are not epineurals because all neoneural-bearing genera have a series of normal percomorph epineurals beneath the neoneurals. The distribution of these ligaments should be studied further. In a preliminary check we have found them (or something like them) in representatives of several other percomorph families: Teraponidae, Kuhliidae, Carangidae, Echeneidae, Lutjanidae, Caesionidae, Gerreidae, Sparidae, Lethrinidae, Sciaenidae, Haemulidae, Labridae, and Scaridae. We have found neoneural ligaments to be absent in representatives of a much larger range of percomorph families. Their presence may be significant, but more work is needed.

In *Bothus* and *Samaris*, neoneurals are ossified in a pattern mimicking epineurals of lower (nonacanthomorph) teleosts (for bothids, see Amaoka, 1969, figs. 120–122, 125). In our *Bothus* specimens, the first bone in the series attaches to the occiput by ligament. The second attaches by ligament to the reduced neural spine of V1, and the next to the neural arch of V2. There is no neoneural bone or ligament on V3 in the two specimens we examined. On V4–6, the bones are forked or spatulate proximally, where they attach to neural arches. On V7–10, the bones have anteromedial branches that attach to the neural arch and unattached anteroventral branches, like the proximally forked epineurals of many lower teleosts (notation ${}_VB_1$ in Tables 3–5). These last seven bones (V4–10) are brush-like distally. Behind V10, the bones are unattached to the axial skeleton and are brush-like both proximally and distally. These unattached bones appear to be serially homologous with the anteroventral branch and posterior body of the bones on V7–10; the posterior epineurals of many lower teleosts show the same pattern (notation V_0 in Tables 3–5). In *Samaris*, the pattern of the neoneurals is similar but differs in detail. The first bone attaches by ligament to the first neural arch, which also (in our specimen) carries a second ligament with no included bone. There are bones attached by ligament to the second and third neural arches, and beyond that a series of unattached bones extends back to V32, some of them are forked proximally, some distally, and some at both ends (Table 8). We interpret this series of bones in *Bothus* and *Samaris* as a neoformation because there is a normal series of percomorph epineurals beneath them, on V2–9 in both genera. Amaoka (1969, figs. 120–122, 125) illustrates variants of the pattern in bothids.

In the section on acanthomorph epineurals and epicentrals, above, we described and commented on possible neopleurals in the polynemid *Polydactylus,* which may be ventral analogues of the neoneurals above the vertebral column.

OCCIPUT.—There are no intermuscular bones or ligaments on the occiput in any acanthomorphs in Tables 7 and 8, except for a ligament on the exoccipital in the stephanoberyciform *Barbourisia* and a neoneural bone attached by ligament to the occiput in the pleuronectiform *Bothus* (there may be two occipital neoneurals in bothids, Amaoka, 1969, figs. 120, 125). One or two intermuscular bones attach to the occiput in some scombroids (e.g., Collette and Chao, 1975).

BAUDELOT'S LIGAMENT.—Among acanthomorphs sampled in Tables 7 and 8, Baudelot's ligament originates on V1 in lampridiforms, *Polymixia,* paracanthopterygians, and stephanoberyciforms and on the occiput in beryciforms, zeiforms, and percomorphs (sensu Johnson and Patterson, 1993). The only exceptions to that statement in Tables 7 and 8 are the two veliferid lampridiforms and the stephanoberyciform *Rondeletia,* where the ligament is double, with part on V1 and part on the occiput (also in the xenisthmid gobioid *Tyson;* Springer, 1988, table 1); the hexagrammid scorpaeniform *Hexagrammus,* where the ligament is on V1; and taxa lacking the ligament, i.e., the beryciforms *Anoplogaster* and *Monocentris,* the four gasterosteiforms, and the gobiesocid *Lepadogaster.* We therefore believe that the point of origin of the ligament is a generally consistent and useful character, and we have used the occipital origin as a character of Euacanthopterygii (Johnson and Patterson, 1993; Percomorpha + Beryciformes and Zeiformes).

RIBS.—Among acanthomorphs sampled in Tables 7 and 8, the primitive teleostean condition (all ribs preformed in cartilage) persists in *Polymixia,* percopsiforms, stephanoberyciforms (*Barbourisia* has the last rib in membrane), several beryciforms (the last rib is in membrane in *Diretmus, Holocentrus, Centroberyx,* anomalopids), mugiloids, atherinomorphs, and various perciforms. All ribs are in membrane in zeiforms (or in ligament), *Elassoma,* gasterosteiforms, *Ammodytes, Lepadogaster,* and gobioids (Johnson and Brothers, 1993).

Ribs show the most remarkable modifications in the pleuronectiforms *Bothus* and *Samaris,* if our interpretation in Table 8 is correct. Amaoka (1969) interpreted bothids as lacking ribs, but Hensley and Ahlstrom (1984) suggested another interpretation, that Amaoka's "abdominal hypomerals" are pleural ribs. In Table 8, we interpret all Amaoka's hypomerals (not just those in the abdominal region) as ribs. According to that interpretation, the first rib of *Bothus* and *Samaris* is on V2, whereas in the other pleuronectiforms studied by Amaoka (1969) it is in the primitive position, on V3. In our *Bothus* specimens, the first three or four ribs are unmodified, but the fourth or fifth and its successors are brush-like distally. The sixth to ninth ribs (on V7–10) have an anterodorsal branch, just like that on the epipleurals of many lower teleosts (Tables 3–5), and from V11 onward the series is continued back into the caudal region by bones that are brush-like at both ends, are unattached to the axial skeleton, and appear to be serially homologous with the anterodorsal branch and posterior body of the bones on V7–10; the posterior epipleurals of many lower teleosts show the same pattern (notation D_0 in Tables 3–5). Amaoka's illustrations of bothids (1969, figs. 120–122, 125) show variants of this pattern. In *Samaris,* the first six ribs (on V2–7) are unmodified, the seventh (V8) is forked proximally, and those on V9–10 have anterodorsal branches passing medial to the proximal parts of the preceding ribs. On V11 (the first caudal) there is a similar bone, with a similar anterodorsal process, but it is not attached to the vertebra. The remaining members of the series are unattached, but on V12–14 the proximal end of the bone passes anterodorsally, medial to the proximal part of the preceding bone, like the anterodorsal branch on the bones on V9–11. The series extends back to V31 (PU9), with some bones forked or bifid distally, some proximally, and some simple. Extra slivers of unattached bone accompany the bones on V10–13. If these bones in the caudal region of bothids and *Samaris* are not ribs, the only alternative interpretation is that they are epipleurals, secondarily redeveloped in a pattern mimicking that in lower (nonacanthomorph) teleosts. Because of the perfect continuity between undoubted and unmodifed anterior ribs, modified posterior ribs, and the unattached bones in the caudal region, we prefer the interpretation that the ribs are modified in an autapomorphic pattern.

In the anabantoid *Colisa* (Table 8) ribs continue back into the caudal region (to PU5) as intermuscular bones, a condition resembling the caudal extension of ribs in bothids.

COMMENTS ON INDIVIDUAL GROUPS AMONG ACANTHOMORPHS

In this section we comment on aspects of the intermuscular bones and ligaments in various acanthomorph groups, which may bear on relationships.

ACANTHOMORPHA.—One feature of the intermusculars supports the monophyly of Acanthomorpha: absence in all acanthomorphs of proximal forking of epineurals and/or epipleurals. We take this proximal forking to be primitive for elopocephalans (above), and although it also is absent in argentinoids, salmonoids, osmeroids, stomiiforms, and most aulopiforms, it is present in myctophiforms, which currently are regarded as the sister group of acanthomorphs (Johnson, 1992).

LAMPRIDIFORMS.—In Table 7 we record the intermusculars

in three lampridiforms, the veliferids *Velifer* and *Metavelifer*, and the lampridid *Lampris* (information on *Lampris* from a partially dissected specimen and radiographs only), which are the three cladistically most primitive genera of this monophyletic group (Olney et al., 1993). These three lampridiforms differ from all other acanthomorphs in having no epineurals in the horizontal septum, and *Velifer* and *Metavelifer* differ from other acanthomorphs in having all epineurals originating on the neural arch (anteriorly) or spine (posteriorly). Like some aulopiforms, the three lampridiforms have the first epineural displaced ventrally relative to its successors (notation b_1 in Tables 5, 7), but that bone is well above the horizontal septum. Furthermore, unlike all other acanthomorphs except *Polymixia*, these lampridiforms also have epicentral ligaments beginning on V1 (*Velifer*, Johnson and Patterson, 1993, fig. 1) or V3 (*Metavelifer, Lampris*). In *Velifer*, where POTs are easily seen in incident light in our 133 mm SL cleared-and-stained specimen, the first POT originates on the parapophysis of V9 and inserts at the tip of the epicentral ligament of V3. In *Polymixia*, the first POT also originates on V9, but it inserts on the epicentral of V5; thus *Velifer* agrees with *Polymixia* and differs from other acanthomorphs in having POTs that originate on abdominal vertebrae and insert on anterior epicentral ligaments. We have seen no epipleural bones or ligaments in lampridiforms, but we note that in veliferids the swimbladder extends behind the abdominal cavity in paired, horn-like processes that occupy the normal position of epipleurals, and that in *Velifer* the series of ribs is continued posteriorly into the caudal region by an extensive series of ligaments (Table 7). A review of other teleosts in our sample shows that in halosaurs and notacanths some (*Halosaurus*) or all (*Notacanthus*) ribs are in ligament (Table 3), in the cypriniform *Hemibarbus* the last eight ribs are attached by ligament (Table 4), in the aulopiform *Scopelarchoides* the last two ribs are in ligament (Table 5), in the beryciform *Anoplogaster* all ribs are attached by ligament (Table 7), and in the exocoetoid *Exocoetus* the last rib is attached by ligament (Table 8), but all these rib-like ligaments are within the abdominal region; in the beryciform *Monocentris* and the zeiforms *Xenolepidichthys* and *Parazen* anterior ribs are in ligament (Table 7). In other words, the situation in *Velifer* is unique among the teleosts that we have sampled, and we surmise that the ligaments behind the ribs represent the missing, but expected, epipleurals.

Thus, lampridiforms are more primitive than all other acanthomorphs in having no epineurals in the horizontal septum, and veliferids are more primitive than other acanthomorphs in having all epineurals originating on the neural arch or spine rather than some on the parapophysis, rib, or centrum. Lampridiforms agree with *Polymixia* in having epicentral ligaments on the first few vertebrae, and in having POTs originating on abdominal vertebrae and inserting on anterior epicentral ligaments. *Velifer* also may have an extensive series of epipleural ligaments. We therefore believe that lampridiforms are the sister group of all other acanthomorphs, *Polymixia* included. Elsewhere (Johnson and Patterson, 1993) we proposed the name Euacanthomorpha for the group comprising nonlampridiform acanthomorphs, and Holacanthopterygii for the Euacanthomorpha minus polymixiiforms; we argued there that polymixiiforms are the sister group of the Holacanthopterygii.

In *Lampris* (Table 7), epineurals of V3–6 (there are no epineural bones behind V6) originate on the rib, not on the neural arch as they do in veliferids; in this *Lampris* agrees with most euacanthomorphs (Tables 7, 8). But in *Lampris*, epicentral ligaments of V3–5 originate on epineural bones (notation L_N in Table 7); that pattern is unique to this genus among our entire sample, and it corroborates the inference that descent of anterior epineurals to ribs occurred independently in *Lampris* and in euacanthomorphs.

PARACANTHOPTERYGIANS.—There are reasons to believe that the Paracanthopterygii are the sister group of the Acanthopterygii (all other holacanthopterygians; Johnson and Patterson, 1993). Those reasons do not include features of paracanthopterygian intermuscular bones and ligaments, which are either uninformative on paracanthopterygian relationships or provide only conflicting evidence. In Table 7 we record the intermusculars in two percopsiforms (*Percopsis, Aphredoderus*) and two gadiforms (*Raniceps, Gaidropsarus*), and Markle (1989:73) summarizes the condition of the ribs and intermuscular bones in a range of paracanthopterygians. Percopsiforms show the pattern that we take to be primitive for the group: a short series of epineural bones (five in *Percopsis*, up to eight in *Aphredoderus*), all lying in the horizontal septum and originating on the rib from V3 onward. There are no epicentral ligaments and no epineural ligaments behind the series of epineural bones. The two gadiforms lack epineurals on V1–2, a synapomorphy of the group (Patterson and Rosen, 1989:19, as "epipleurals"; there is a small, unattached bone on one side of V1 in our *Raniceps*), but they have a long series of epineurals that extend back into the caudal region. Again there are no epicentral ligaments, and we have not succeeded in finding either epicentral ligaments or POTs in any paracanthopterygians we checked, including ophidiiforms, batrachoidiforms, and lophiiforms.

STEPHANOBERYCIFORMES.—In Table 7 we record the intermusculars of six stephanoberyciforms, representing six of eight families currently included in the group (the Stephanoberycoidei of Moore, 1993). There are from 0 (*Stephanoberyx*) to 10 (*Rondeletia*) epineural bones in stephanoberyciforms (melamphaids have 4–6), and those on rib-bearing vertebrae (V3 onward) originate on the parapophysis or centrum (Table 7), except in *Barbourisia*, where there are epineurals on the ribs of V3–5 (Table 7). In having epineurals of rib-bearing vertebrae on the parapophysis or centrum rather than the rib, stephanoberyciforms resemble *Polymixia* and differ from beryciforms and percomorphs, which have epineurals on the rib on anterior rib-bearing vertebrae. The major exception to the latter condition to be found in Tables 7 and 8 is that all epineurals of rib-bearing vertebrae are on the parapophysis in the atherinoid *Menidia* and the exocoetoid *Exocoetus*. We take this to be secondary because several anterior epineurals are on the rib in the atherinoid *Bedotia* and in mugiloids, gasterosteids, and *Elassoma*, taxa that include outgroups of *Menidia* and *Exocoetus* in our opinion (Johnson and Patterson, 1993). Other exceptions in Tables 7 and 8, such as the zeiform *Xenolepidichthys*, with ribs formed only in ligament, and *Cottus*, with ribs only on V9–11, are clearly secondary. We therefore take the origin of epineurals on the parapophysis in stephanoberyciforms to be a primitive condition, differentiating them from beryciforms and percomorphs (Johnson and Patterson, 1993). We interpret the descent of epineurals to the ribs in *Barbourisia* to be independent of their descent in beryciforms and percomorphs, given the cladistically derived position of *Barbourisia* in Moore's (1993) phylogeny of stephanoberyciforms (placed as the sister group of megalomycterids + cetomimids). *Lampris* (above) and paracanthopterygians provide other examples of independent descent of epineurals to ribs.

We have not found epicentral ligaments in any stephanoberyciform. We agree with Moore (1993) that melamphaids are the most primitive stephanoberyciforms, and they are the only members of the group in which we have found POTs; the first POT originates on V9 in *Poromitra*, on V10 in *Scopelogadus*, and on V12 in *Melamphaes*. Because there are no epicentral ligaments, the POTs of melamphaids do not cross any other ligaments in their passage to the lateralis musculature, and the anterior ones do not insert on epineurals.

Rondeletia is different from any other stephanoberyciform in our sample in several ways. It has epineural bones on V1–10, more than any other stephanoberyciform. All epineural bones lie in the horizontal septum, and the epineurals are continued caudally by a series of ligaments (V11–16) in the horizontal septum, originating on the haemal arch anteriorly and the centrum posteriorly. Finally, at the tip of the epineurals on V1–8 there are small, superficial plates of cartilage, like those in *Holocentrus* (Figure 13). Given the position of *Rondeletia* in Moore's (1993) phylogeny of stephanoberyciforms (as the sister group of Barbourisiidae and Megalomycteridae + Cetomimidae), we interpret these features of its intermusculars as autapomorphous, not primitive for the group (the features are all absent in melamphaids and *Gibberichthys*, for example).

EUACANTHOPTERYGII.—Elsewhere (Johnson and Patterson, 1993) we proposed the name Euacanthopterygii for a group comprising Beryciformes and an expanded Percomorpha (including Atherinomorpha). The only character of the intermusculars bearing on that grouping is the origin of epineurals on ribs (rather than on the parapophysis, centrum, or neural arch) of anterior rib-bearing centra. The character is questionable because the same origin of epineurals on the ribs is found in the lampridiform *Lampris*, the stephanoberyciform *Barbourisia*, and in paracanthopterygians.

ZEIFORMES.—Elsewhere (Johnson and Patterson, 1993) we argued that Zeiformes do not include caproids (*Capros*, *Antigonia*), and that caproids are percomorphs but zeiforms are not. As with paracanthopterygians (above), the intermusculars of zeiforms are uninformative or provide contradictory information on zeiform relationships. In Table 7 we record intermusculars of the grammicolepid *Xenolepidichthys* and the parazenid *Parazen*, two genera that we regard as relatively primitive zeiforms (Johnson and Patterson, 1993). *Xenolepidichthys* has epineural bones on V2–7 and a ligament on V8. There is then a gap of three vertebrae before a series of epicentral ligaments on the haemal arches of V12–20. None of the epineurals originates on a rib, but "ribs" are in ligament on V4–8, where they overlap the epineural series. *Parazen* shows a unique condition among the acanthomorphs in Tables 7 and 8; epineural bones are present on V1–13, but they are all unattached except for those on V5–6, which originate on the ligamentous ribs of those vertebrae. Unattached epineurals occur commonly in zeiforms (e.g., *Cyttopsis*, *Zenion*).

We interpret the separation of the epineural and epicentral series in *Xenolepidichthys* as relatively primitive, resembling the condition in centropomid percoids and in mugiloids. But absence of overlap between the epineural and epicentral series, and the fact that the epineurals have not ascended above the haemal arch by V8, give no support to the proposal by Johnson and Patterson (1993) that zeiforms fit between stephanoberyciforms and beryciforms in the acanthomorph cladogram, for beryciforms show a more primitive state than zeiforms in both these characters (Table 7). The POTs of zeiforms also show a pattern more like that in percomorphs than beryciforms, with some anterior POTs originating on abdominal vertebrae and inserting on epineural bones (as noted above, in *Parazen* the first POT originates on V12 and inserts on the epineural of V3, and in *Xenolepidichthys* the first POT originates on V6 and inserts on the epineural of V2).

BERYCIFORMES.—Table 7 records the intermusculars of nine beryciforms, including representatives of all seven families (unlike Stiassny and Moore, 1992, and Moore, 1993, we include the Berycidae and Holocentridae in Beryciformes; Johnson and Patterson, 1993). We take the pattern in holocentrids, described in detail above and illustrated in Figures 12 and 13, to be primitive for the group. Major components of that pattern include a series of epineural bones that originates on the ribs on anterior rib-bearing centra and ascends to the neural spine on posterior abdominal and caudal vertebrae and a series of epicentral ligaments beginning at about the first caudal vertebra. There are epipleural ligaments on the first two caudal vertebrae in holocentrids, but we have not seen similar structures in other beryciforms.

The holocentrid pattern of intermusculars is most closely matched in *Anomalops* and the berycid *Centroberyx* (Table 7). In the anomalopid *Photoblepharon* and in *Trachichthys* the pattern is similar, but the epineural series is interrupted, with bones on V1–2 and ligaments (with included bone posteriorly in *Photoblepharon*) from V6 or V7. *Monocentris* lacks epineurals on V1–7 and ribs on V3–4, and its posterior epineurals (identifiable as epineurals because there are epicentral ligaments beneath them) remain in the horizontal septum, rather than ascending on to the neural arch and spine as they do in other beryciforms. In *Diretmus* there is only one epineural, on V1, and in *Anoplogaster,* which appears to be the sister group of diretmids (Moore, 1993; Baldwin and Johnson, in press), there are no epineurals or epicentrals, no Baudelot's ligament, and all ribs are attached by ligament.

PERCOMORPHA (sensu Johnson and Patterson, 1993).—The Percomorpha may be rendered monophyletic by inclusion of the Atherinomorpha and exclusion of several groups (Lampridiformes, Stephanoberyciformes, Beryciformes, Zeiformes) (Johnson and Patterson, 1993). The Percomorpha in that sense share three derived features of the intermusculars (Table 8): (1) all epineurals lie in the horizontal septum, originating on the ribs anteriorly (V3 onward) and on the parapophyses, haemal arches, or centra posteriorly; (2) the epicentral series of ligaments, when present and distinct, never overlaps the epineural series; and (3) POTs in the abdominal region extend anteriorly beyond the epicentrals to insert on epineural bones. As with so many other characters in percomorphs and their relatives, the percomorph condition of all three characters occurs elsewhere in acanthomorphs. Regarding character (1), epineurals show the same pattern in paracanthopterygians (e.g., gadiforms, Table 7), in the stephanoberyciform *Rondeletia,* and in zeiforms. Regarding character (2), the zeiform *Xenolepidichthys* shows no overlap between epineurals and epicentrals (Table 7). And regarding character (3), zeiforms show the percomorph pattern.

The pattern of intermusculars that we interpret as primitive for percomorphs is that illustrated in Figure 14 and shown in Table 8 by centropomids, *Morone, Drepane,* larval *Sphyraena,* and mugiloids. In that pattern, the epineural and epicentral series are discrete, with a gap of from one (*Psammoperca*) to seven (*Lates calcarifer*) vertebrae between the last epineural and the first epicentral. In many percomorphs the epineural and epicentral series become indistinguishable (e.g., caproids, *Siganus, Colisa, Rachycentron, Menidia;* Table 8). The problem of distinguishing the two, the possible value of ontogeny in doing so, and our usage in Table 8 are discussed above (p. 38).

We believe that a wider and more thorough survey of percomorph intermusculars might well provide valuable information on relationships. As one example, the neoneural series of ligaments, recorded in Table 8 in the ammodytoid *Ammodytes,* the polynemoid *Polydactylus,* and the mullid *Upeneus* and discussed above (p. 41), may prove to be a useful feature when its distribution is better understood. Two further examples concern gobieseocids and uranoscopids. In Table 8, intermusculars of the gobiesocid *Lepadogaster* are recorded. They show a remarkable pattern, with all epineurals of rib-bearing vertebrae (V2–14) originating on the centrum, and with ribs originating on the epineural (notation B_N) on V2–11 and lying free on V12–14. This pattern (ribs originating on epineurals) is evidently general in gobiesocids (Gosline, 1970:376) but does not occur in callionymids (*Callionymus,* Table 8), which are their close relatives (Winterbottom, 1993; Johnson and Patterson, 1993:614). Callionymids lack ribs, but they share with gobiesocids absence of an epineural on V1, an unusual feature (caproids and the zeiform *Xenolepidichthys* are the only other taxa in Tables 7 and 8 that lack the epineural on V1 but have one on V2).

Concerning uranoscopids, Figure 16 is a reinterpretation of Pietsch's (1989, fig. 31) drawing of the axial skeleton of *Uranoscopus.* In the illustrated specimen of *Uranoscopus* there are ribs on V3–12 and epineurals on V1–17. Pietsch (1989) interpreted the first four unattached epineurals as ribs and the first three ribs as epineurals (his epipleurals); based on our reinterpretation, he counted 10 pairs of ribs, missing the small ones of V12, and 15 pairs of intermusculars, missing the short epineural of V1. That bone is variable, being present only on the right side in the illustrated specimen, present on both sides in a second specimen and in that used by Starks (1923:279), and absent on both sides in a third. In the illustrated specimen, epineurals of V1–5 are unattached, and those of V6–17 originate on the parapophysis (anteriorly) or centrum (posteriorly). The ribs of V3–5 originate on the parapophysis in the normal way, but those of V6–12 are unattached, resembling the epineurals of V2–5 in form and orientation, but distinguishable from those by their position beneath the epineurals and by their cartilaginous tips. We have examined three other genera of uranoscopids. *Gnathagnus* has three attached ribs, so that the switch between attached ribs and attached epineurals is at V6, as in *Uranoscopus. Kathetostoma* has the first epineural associated with V2 and has the first four ribs attached, so that the switch is at V7. *Astroscopus* has the first epineurals associated with V1 but is otherwise like *Kathetostoma,* with the switch at V7. Okamura and Kishimoto (1993) described a new uranoscopid genus, *Selenoscopus,* and illustrated V9–14 in it and *Uranoscopus.* They described *Selenoscopus* as having nine pairs of pleural ribs and "epicentrals" and 10 pairs of epipleurals. Comparison of their illustrations with Figure 16 indicates that their "epipleurals" on V9–11 are ribs, their "pleural ribs" on V9–11 are epineurals, and their "epicentrals" are the epineurals of caudal vertebrae.

In Pietsch's (1989) cladogram, trachinids are the sister group of uranoscopids. *Trachinus* (Table 8) has epineurals from V1, all attached to the axial skeleton. There are only four ossified ribs, on V10–13, but there are ligaments in the position of ribs on several preceding vertebrae. The sister group of trachinids + uranoscopids in Pietsch's (1989) cladogram is the leptoscopids, whereas in Pietsch and Zabetian's (1990) cladogram it is ammodytids. In *Leptoscopus* there are no ribs, and epineurals extend from V2 back into the caudal region, all attached to the axial skeleton. In *Ammodytes* (Table 8) there are epineurals attached to the axial skeleton from V1 back into the caudal region, normal ribs, and a conspicuous series of neoneural ligaments from V5 back into the caudal region. The

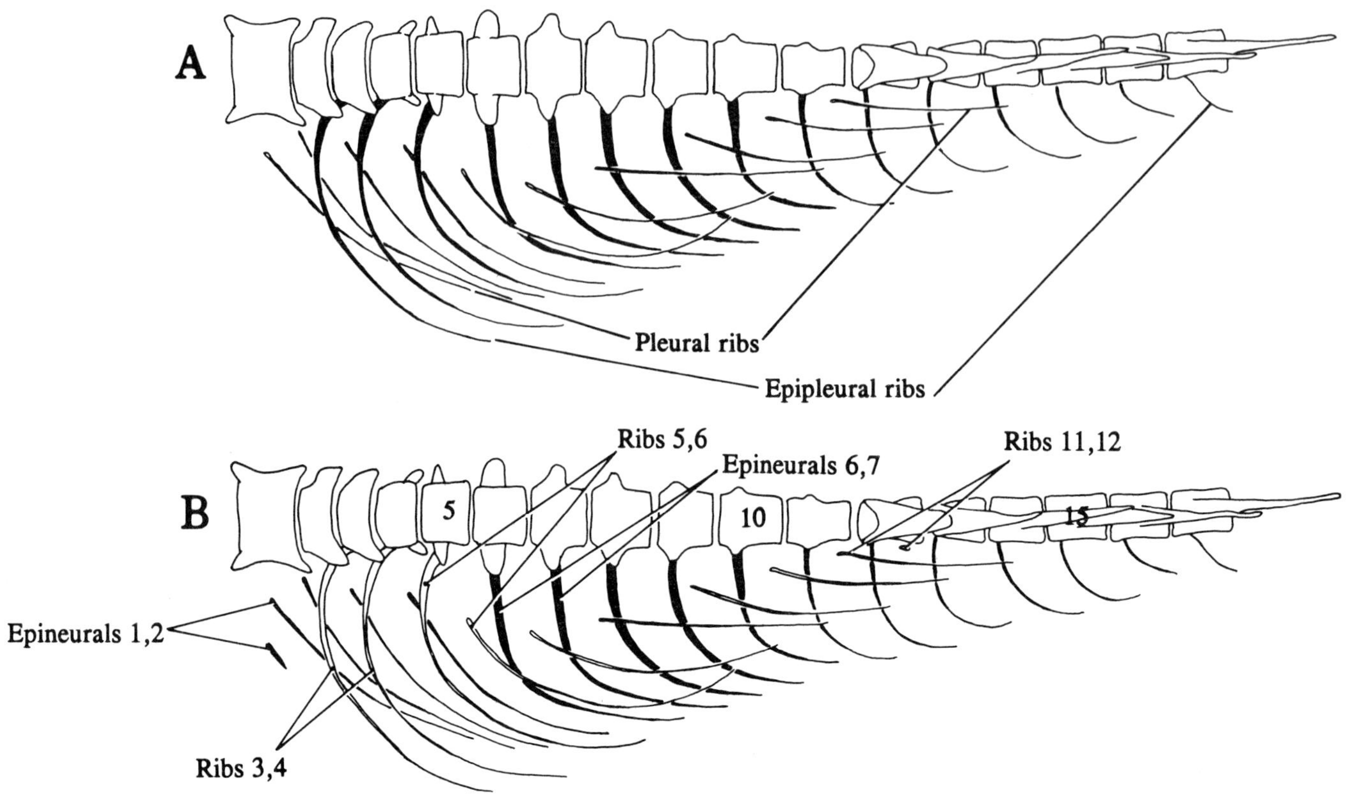

FIGURE 16.—Ribs and intermuscular bones of *Uranoscopus scaber* L. A, "Semi-diagrammatic view of axial skeleton of ... USNM 206551, 75 mm; ventral view, showing pleural and epipleural ribs of right side" as interpreted by Pietsch (redrawn from Pietsch, 1989, fig. 31, modified by emphasizing bones that he interpreted as intermusculars in solid black). B, Same drawing reinterpreted, with epineurals shown in solid black and epineural of V1 (present only on one side of specimen) and diminutive rib of V12 added. Every fifth vertebra is numbered.

next group in Pietsch's (1989) and Pietsch and Zabetian's (1990) cladograms is Champsodontidae + Chiasmodontidae. *Champsodon* has a normal percoid pattern, with epineurals beginning on V1 and descending to the ribs from V3 onward. In *Chiasmodon*, the first epineural is on V1, and from V3 the rib and epineural are fused proximally at their common articulation with the parapophysis, and the body of the rib and epineural have almost the same horizontal orientation, diverging only slightly distally. In *Dysalotus* the pattern is the same except that the rib and epineural are not fused proximally.

We conclude from this that uranoscopids are characterized by a distinctive pattern, the switch between attached ribs and attached epineurals (Figure 16), and that the pattern of ribs and epineurals is considerably more variable among Pietsch's trachinoids than in any other percomorph group. We think it likely that further study of this complex will challenge the hypothesized monophyly and relationships of the Trachinoidei.

SMEGMAMORPHA.—The Smegmamorpha (Johnson and Patterson, 1993) comprise the Synbranchoidei (synbranchids and mastacembeloids), Mugilomorpha (of Stiassny, 1990, = Mugiloidei), Elassomatidae (*Elassoma* only), Gasterosteiformes (including *Pegasus* and *Indostomus*), and Atherinomorpha. Table 8 records the intermusculars in two mugiloids, *Elassoma,* three atherinomorphs, and four gasterosteiforms. The principal character of smegmamorphs is to have the first (or first and second) epineural originating at the tip of a lateral process from the first (or first and second) vertebra (notation B_P in Table 8). To our knowledge, this character occurs in only two other teleosts, the carangoid *Echeneis* and the gobiid *Gnatholepis*. As *Echeneis* and *Gnatholepis* are each nested well within well-characterized monophyletic groups (carangoids, Johnson, 1984; gobioids, Winterbottom, 1993), we interpret the modified first epineural origin in them as independently derived.

PLEURONECTIFORMES.—We recorded the intermusculars in only two genera of pleuronectiforms, the bothid *Bothus* and the samarid *Samaris*. We chose those genera because bothids and samarids are known to have "extra" series of intermusculars (Hensley and Ahlstrom, 1984). According to our interpretation, *Bothus* (and other bothids, Amaoka, 1969) and *Samaris* uniquely share two derived features. The first is ossification of the neoneurals, which, to our knowledge, does not occur elsewhere in teleosts. The second is modification of the posterior ribs so that they come to resemble the epipleurals of lower teleosts and extend into the caudal region as intermuscular bones (see section on "Ribs," above). In our sample, caudal extension of ribs occurs elsewhere only in the anabantoid *Colisa* (Table 8), where the bones do not show the proximal branching characteristic of the posterior ribs in bothids and *Samaris*. In addition, *Bothus* (and other bothids, Amaoka, 1969, as "hypomerals") share with *Samaris* the derived condition of origin of the first rib on V2, rather than on V3, the primitive position, as in the other pleuronectiforms studied by Amaoka (1969).

Pleuronectiform interrelationships are still poorly resolved (Chapleau, 1993). Bothids and samarids are separated by two nodes in Chapleau's (1993) cladogram, with bothids in a polychotomy together with several other lineages, and samarids grouped with soleoids. The characters cited above imply that bothids and samarids are immediately related, and this should be taken into account in future studies of pleuronectiform interrelationships.

Conclusions and Summary of Systematic Consequences

We believe that the intermuscular bones and ligaments of teleosts have been neglected and are potentially of value in systematics. As an analogy, the teleostean caudal skeleton generally was neglected as a source of characters until the 1960s; for example, Gosline (1960:328) wrote that it was "usually either ignored or briefly dismissed in taxonomic papers." The reason for that neglect was put nicely by Monod (1968:5): "Authors coming to the caudal skeleton in their osteological descriptions find themselves a little embarrassed or ill at ease because of the unfamiliar ground under their feet: is the ice, just here, enough to bear my weight? All those who have grappled with this part of the skeleton will have experienced the sensation to some extent" (our loose translation from the French). One of us can vouch for the truth of Monod's diagnosis in the 1950s and early 1960s, and the tentative and sketchy references to the caudal skeleton in Greenwood et al. (1966) confirm it. Yet today every ichthyologist is familiar with the caudal skeleton, and many details of it are widely used in systematics. Again, Monod (1968:5) correctly diagnosed one reason for that change: "The major obstacle ... is above all the lack of a coherent and homogeneous terminology." The second reason for the change was the lack (before Monod, 1968) of a broad comparative survey of the teleostean caudal skeleton. Once the kinds of variations in the caudal skeleton were mapped, and nomenclature was at hand to describe them, ichthyologists were happy, in Monod's simile, to venture on to the ice.

As one example of the current confusion over teleostean intermusculars we cite Whitehead and Teugel's (1985) account of the clupeoid *Sierrathrissa*, surely one of the best anatomical papers of the last decade. Yet when they came to the intermusculars, Whitehead and Teugels were lost; they called the epineurals by two different names (epicentrals, dorsal intermuscular bones), which were both wrong; they wrongly called the epicentrals epipleurals; and they wrongly called the epipleurals ventral intermuscular bones—the ice would not support them. These errors are no worse than those of earlier generations, as when Goodrich (1909, fig. 305; 1930, fig. 82), in adapting two figures of *Salmo* from Bruch (1861), converted the epicentral ligaments and cartilages into solid bones, which he called ribs, and wrongly called the epineurals epipleurals.

We do not pretend that this paper stands with Monod's (1968) monograph on the teleostean caudal skeleton. (Monod admitted that although he spent almost 10 years on that work, he still felt that he should have studied a hundred times as many species and 10 times as many specimens per species (1968:3); we feel the same about our more cursory survey of teleostean intermusculars.) Nevertheless, we hope that the information on variation in the intermusculars and the terminology that we have provided are enough to convince others that the ice will bear their weight, and that it is worth the effort of learning to read the intermuscular bones and ligaments.

By way of a concluding summary, we first describe the system of intermuscular bones and ligaments that we believe to be primitive for Recent teleosts, and then we summarise modifications of that system that characterize various major groups.

The primitive condition of the intermusculars is exemplified by *Hiodon*. The ossified epineurals are all fused with the neural arch proximally, and the epineural series is extended into the caudal region by about a dozen ligaments. Epicentrals are present only as ligaments, originate on the parapophysis on rib-bearing vertebrae, and extend into the caudal region where their origin ascends to the centrum. Epipleurals also are present only as ligaments, and they are confined to the middle of the body, where they originate on the last few ribs and the first few haemal arches.

Within osteoglossomorphs, there are many variations on the basic pattern seen in *Hiodon* (Taverne, 1977b, 1978, and above), but our survey has not been complete enough to determine the systematic significance of those variations. Elopocephalans (nonosteoglossomorph teleosts) generally share two features that are not found in any osteoglossomorph: a long series of epipleural bones and proximal forking of many of the epineural and epipleural bones. Fusion (continuity) between the epineural and the neural arch, true of all epineurals in *Hiodon* and of all but the first in *Pantodon*, persists on anterior vertebrae in many lower elopocephalans (Tables 3–5); epineurals fused to the neural arch never show proximal forking. There is sometimes (and probably primitively) an abrupt switch from a fused, simple epineural to a free, proximally forked one on the succeeding vertebra (*Elops*, *Megalops*); *Clupea*, *Chirocentrus*, and *Esox lucius* show a more gradual transition between the two conditions, over two to four vertebrae. The presence of epicentral bones in all clupeomorphs (including *Denticeps*) and all gonorynchiforms provides additional support for the monophyly of each of those groups. Derived fusion between anterior epicentrals and ribs in clupeids and chirocentrids supports the monophyly of that grouping, but it is incongruent with what we know about the distribution of cartilage chevrons at the tips of the anterior epicentrals, which implies a grouping of engraulids and clupeids. More double-stained material is necessary to resolve the conflict.

We take the presence of epipleural bones and of proximally forked epineurals and epipleurals to be primitive for elopocephalans because of the conditions in elopomorphs, clupeomorphs, *Esox,* and gonorynchiforms. The second of those characters is absent in argentinoids, salmonoids, and osmeroids, and that absence is conceivably a character of a group including only those taxa (Salmoniformes). Intermusculars provide stronger support for a relationship between salmonoids and osmeroids (sensu Begle, 1991, 1992) because those groups share absence of epipleural bones and presence of cartilaginous rods in the epicentral ligaments.

The intermusculars show remarkable variations in aulopiforms, and we use characters from them as the basis of a parsimony analysis. Results of that analysis (Figure 11) suggest some radical reinterpretations of aulopiform interrelationships.

In acanthomorphs, features of the intermusculars support monophyly of a group comprising all acanthomorphs except lampridiforms (Euacanthomorpha, defined by descent of one or more epineurals into the horizontal septum); one comprising all euacanthomorphs except polymixiids (Holacanthopterygii, defined by descent of several or all the epineurals into the horizontal septum and absence of epipleural bones); one comprising zeiforms and Percomorpha sensu Johnson and Patterson (1993; the characters are an association between POTs and epineural bones and absence of overlap between the epineural and epicentral series); and one comprising synbranchoids, mugiloids, *Elassoma,* gasterosteiforms, and atherinomorphs (Smegmamorpha, defined by origin of the first epineural on a transverse process of V1). There is weaker support from the intermusculars (weaker in the sense of involving more homoplasy in outgroups) for the monophyly of Acanthomorpha (absence of proximal forking of epineurals and/or epipleurals) and of Euacanthopterygii (Acanthopterygii minus stephanoberyciforms, defined by origin of anterior epineurals from the rib rather than the parapophysis).

Much work remains to be done on the intermuscular bones and ligaments, particularly within percomorphs, where features like the neoneural ligaments that we report in a few taxa and the possible neopleurals in *Polydactylus* suggest characters that may be of value in systematics.

Finally, there remains the question of whether it is possible to provide efficient anatomical definitions or diagnoses of the different series of intermusculars. In naming the epineural, epicentral, and epipleural series "according to the vertebral element they may adhere to," Owen (1866:43) noted that this criterion failed because "each may shift its place, rising or falling gradually along the series of vertebrae." When the intermuscular system is unmodified, the horizontal septum is the best criterion: epicentrals are in the horizontal septum, epineurals are above it, and epipleurals are below it. POTs, when present, are always in the horizontal septum, and their association with the epicentrals (AOTs) provides an additional guideline in the unmodified system.

A second criterion is developmental sequence. Epineurals and epicentrals develop in a rostrocaudal gradient, whereas epipleurals develop both rostrally and caudally from a focus around the first caudal vertebra. We know of no exception to the first part of that statement (concerning epineurals and epicentrals), although both series may be truncated rostrally. For example, the first epineural is on about V20 in *Arapaima,* V12 in *Denticeps,* V10 in *Monocentris,* V3–4 in *Salangichthys* and *Stokellia,* and V3 in *Jenkinsia* and gadiforms; the first epicentral is on about V15–20 in synodontids, *Saurida, Pseudotrichonotus, Chlorophthalmus,* and notosudids among aulopiforms, on about V5 in the aulopiforms *Coccorella* and *Harpadon,* on about V7–10 in *Diretmus, Monocentris,* trachichtyids, anomalopids, berycids and holocentrids among beryciforms, and in mugilids, centropomids, *Sphyraena,* and *Polydactylus* among percomorphs, and behind V10 in zeiforms and in *Morone, Doederleinia,* and *Drepane* among percomorphs. Ontogenetic information is lacking in many of these taxa with rostrally truncated epineural or epicentral series, and in those cases rostrocaudal development remains an untested assumption. Bidirectional development of the epipleural series is known to fail as a criterion in one group, aulopiforms, where we have limited developmental information, and it may fail in another group, alepocephaloids, where we lack that information.

The horizontal septum fails as a defining criterion in two taxa, aulopiforms and nonlampridiform acanthomorphs (Euacanthomorpha of Johnson and Patterson, 1993). In many aulopiforms, according to our interpretation, the anterior epipleurals lie in the horizontal septum, and in euacanthomorphs some (*Polymixia,* beryciforms) or all (paracanthopterygians, stephanoberyciforms, zeiforms, percomorphs) of the epineurals are in the horizontal septum. The attachment of POTs to epicentrals also fails as a criterion in zeiforms and percomorphs, where the POTs may attach to descended epineurals.

The problem of defining or diagnosing the intermuscular series in these modified groups is similar to the problem of deciding whether the ribs of a gnathostome are dorsal ribs or ventral (pleural) ribs (Rosen et al., 1981:242). With ribs, topographic criteria fail because the ribs may "wander" within the myoseptum from one part of the vertebral column to another in an individual or from species to species, and because the horizontal septum is absent in most tetrapods. According to our interpretations, the epineural and epipleural intermusculars also "wander" within the myoseptum and so come to lie in the horizontal septum in aulopiforms and euacanthopterygians. Because of these modifications, we are unable to give diagnoses or definitions of the epineural and epicentral series that will apply in all teleosts; in difficult cases, where one or more intermuscular series is lost or modified, identification of the remaining series will have to rely on comparative arguments, assessing details of structure and distribution within a cladistic framework.

Tables

Key to Tables 1–5, 7, 8

Headings of rows and columns:

- ANA — Accessory neural arch
- Ec — Epicentral series
- En — Epineural series
- Ep — Epipleural series
- Nn — "Neoneural" series in certain percomorphs (Table 8)
- Occ — Occiput
- R — Pleural ribs
- 1,2, etc. — vertebrae, numbered from the first
- // — Symbols separate vertebral numbers to indicate that a series of vertebrae is ommitted (intermuscular elements unvaried in that series)

Entries in rows and columns:

- B — Bone (intermuscular or rib) not preformed in cartilage (membrane bone)
- B (underlined) — Rib preformed in cartilage (cartilage bone)
- *B* — Bone (intermuscular or rib) attached to axial skeleton by ligament
- **B** — Epineural bone displaced ventrally into horizontal septum (acanthomorphs, Tables 6, 7) or epipleural bone displaced dorsally into horizontal septum (aulopiforms, Table 5)
- ***B*** — As **B** above, but attached to axial skeleton by ligament
- b — Anterior epineural bone placed lower (lower case) distally than its successors, but above the horizontal septum
- C — Rib in cartilage (unossified)
- D — Epipleural bone, a member of a series with anterodorsal branching, in which the medial branch, attaching to the axial skeleton, is lost, so that the remaining part represents only the anterodorsal branch and the posterior part of the bone
- F — Intermuscular bone or rib fused to axial skeleton; epineural bones are primitively fused to the neural arches, epicentral bones are occasionally fused to ribs, and anterior ribs are rarely fused to parapophyses (*Esox*) or to centra (*Alepisaurus, Paralepis*)
- **F** — Epineural bone displaced ventrally into horizontal septum and fused with a rib (*Dascyllus*, Table 8)
- f — Anterior epineural bone fused to axial skeleton and placed lower (lower case) distally than its successors, but above the horizontal septum
- L — Ligament attached to axial skeleton
- L (underlined) — Epicentral ligament containing a rod of cartilage distally
- **L** — Epineural ligament displaced ventrally into horizontal septum
- M — Myorhabdoi: unattached, superficial bones succeeding the epineurals and in the epipleural position in *Xenomystus* (Table 3)
- **PU** — Number of preural centrum following the last element of the intermuscular system
- **S** — Baudelot's ligament, originating on the occiput or first vertebra and inserting on the supracleithrum
- *S* — Ossified Baudelot's ligament
- V — Epineural bone, a member of a series with anteroventral branching, in which the medial branch, attaching to the axial skeleton, is lost, so that the remaining part represents only the anteroventral branch and the posterior part of the bone
- X — "Cranial ribs" on occiput and V1 of *Gonorynchus*
- ?> — In *Lampris* (Table 7), conditions unknown posteriorly, partially dissected specimen only
- |, /, ___ — Fusion between elements connected or indicated by the symbol

Subscripts

As prefix:

- C — With a chevron of cartilage distally
- D — With a dorsal branch anteriorly
- P — Epicentral ligament reduced to a short "sling" running from the centrum or parapophysis to the upper surface of an epipleural bone
- V — With a ventral branch anteriorly
- < — Branched or forked distally
- > — Branched or forked proximally
- X — Branched or forked at both ends (proximally and distally)

As suffix:

- 0 — Unattached to axial skeleton
- 1 — Attached to neural arch or spine
- 2 — Attached to centrum
- 3 — Attached to parapophysis
- 4 — Attached to rib
- 5 — Attached to haemal arch or spine
- E — Attached to epipleural bone
- N — Attached to epineural bone
- P — Attached to lateral process of vertebra

TABLE 1.—Distribution of intermuscular bones and ligaments in *Polymixia*.

		Occ	ANA	1	2	3	4	5	6	7	8	9	10	11	12	13	14	15	16	17	18	19	20	21	22	23	24	25	26	27	28	29	30	31
Polymixia	En	-		B_1	B_1	B_3	B_3	B_3	B_3	B_3	B_3	B_2	B_2	B_1	B_1	B_1	B_1	B_1	B_1	B_1	B_1	B_1	B_1	L_1	L_1	L_1	L_1							
lowei	Ec				L_1	$\underline{L}_4$	$\underline{L}_4$	$\underline{L}_4$	$\underline{L}_4$	$\underline{L}_4$	$\underline{L}_4$	$\underline{L}_3$	$\underline{L}_3$	$\underline{L}_3$	$\underline{L}_3$	$\underline{L}_5$	$\underline{L}_5$	$\underline{L}_5$	$\underline{L}_5$	$\underline{L}_5$	L_5	L_2	L_2	L_2	L_2	L_2	L_2	**PU4**						
BM 87.12.7.1	Ep										L_4	L_4	B_4	B_4	B_4	B_5	B_5	B_5	B_5	B_5	B_5	B_5	L_5	L_5	L_5									
SL 128 mm	R			S		$\underline{B}$	$\underline{B}$	$\underline{B}$	$\underline{B}$	$\underline{B}$	$\underline{B}$	$\underline{B}$	$\underline{B}$	$\underline{B}$	$\underline{B}$																			
Polymixia	En	-		B_1	B_1	B_3	B_3	B_3	B_3	B_3	B_3	B_2	B_2	B_1	B_1	B_1	B_1	B_1	B_1	B_1	B_1	B_1	L_1											
lowei	Ec				L_1	$\underline{L}_4$	$\underline{L}_4$	$\underline{L}_4$	L_4	L_4	L_4	L_4	L_3	L_3	L_3	L_5	L_5	L_5	L_5	L_5	L_2	L_2	L_2	L_2	L_2	L_2	L_2	**PU4**						
USNM 308378	Ep										L_4	B_4	B_4	B_4	B_4	B_5	B_5	B_5	B_5	B_5	B_5	B_5	L_5											
SL 72 mm	R			S		$\underline{B}$	$\underline{B}$	$\underline{B}$	$\underline{B}$	$\underline{B}$	$\underline{B}$	$\underline{B}$	$\underline{B}$	$\underline{B}$	$\underline{B}$																			
Polymixia	En	-		B_1	B_1	B_3	B_3	B_3	B_3	B_3	B_3	B_2	B_2	B_1	B_1	B_1	B_1	B_1	B_1	B_1	L_1	L_1	L_1	L_1	L_1									
lowei	Ec				L_2	L_4	L_4	L_4	L_4	L_4	L_4	L_4	L_3	L_3	L_3	L_5	L_5	L_5	L_2	L_2	L_2	L_2	L_2	L_2	L_2	**PU6**								
MCZ 95714	Ep											L_4	B_4	B_4	B_4	B_5	B_5	B_5	B_5	B_5	B_5	L_5	L_5											
SL 28 mm	R			S		$\underline{B}$	$\underline{B}$	$\underline{B}$	$\underline{B}$	$\underline{B}$	$\underline{B}$	$\underline{B}$	$\underline{B}$	$\underline{B}$	$\underline{B}$																			
Polymixia	En	-		B_1	B_1	B_3	B_3	B_3	B_3	L_3	L_3	L_2	L_2	L_1	L_1	L_1	L_1	L_1	L_1	L_1														
lowei	Ec																						**PU9**											
MCZ 64773	Ep											B_4	B_4	B_4	B_4	L_5	L_5	L_5	L_5	L_5	L_5	L_5												
SL 12 mm	R					$\underline{B}$	$\underline{B}$	$\underline{B}$	$\underline{B}$	$\underline{B}$	$\underline{B}$	$\underline{B}$	$\underline{B}$	$\underline{B}$	$\underline{B}$																			

TABLE 2.—Distribution of intermuscular bones and ligaments in *Gonorynchus*.

		Occ	ANA	1	2	3	4	5	6	7	8	9	10	11	12	13	14	15	16	17	18	19	20	21	22	23	24	25	26	27	28	29	30	31
Gonorynchus	En	BBB	-	V_0	V_0	L_2V_0	B_2V_0	vB_1	vB_1	vB_1	vB_1	vB_1	vB_1	vB_1	vB_1	vB_1	vB_1	vB_1	vB_1	vB_1	vB_1	vB_1	vB_1	vB_1	vB_1	vB_1	vB_1	vB_1	vB_1	vB_1	vB_1	vB_1	vB_1	vB_1
gonorynchus	Ec	BBB		$<B_3$	B_3		B_3	B_3	B_3	B_3	B_3	B_3	B_3	B_3	B_3	B_3	B_3	B_3	B_3	B_3	B_3	B_3	B_3	B_3	B_3	B_3	B_3	B_3	B_3	B_3	B_3	B_3	B_3	B_3
SIO 70-275	Ep													B_3	B_3	B_3	B_3	B_3	B_3	B_3	B_3	B_3	B_3	B_3	B_3	B_3	B_3	B_3	B_3	B_3	B_3	B_3	B_3D_0	B_3D_0
SL 146 mm	R	XX		X		<u>B</u>	B	B	B	B	B	B	B	B	B	B	B	B	B	B	B	B	B	B	B	B	B	B	B	B	B	B	B	B

		32	33	34	35	36	37	38	39	40	41	42	43	44	45	46	47	48	49	50	51	52	53	54	55
Gonorynchus	En	vB_1	vB_1	vB_1	vB_1	B_1V_0	B_1V_0	B_1V_0	B_1V_0	B_1V_0	L_1V_0	V_0	V_0	V_0	V_0	V_0	V_0								
gonorynchus	Ec	B_3	B_3	B_3	B_3	B_3	B_3	B_3	B_3	B_3	L_3	L_3	L_5	L_2	L_2	L_2	L_2	L_2	L_2	L_2			**PU3**		
SIO 70-275	Ep	B_3D_0	B_3D_0	B_3D_0	B_3D_0	D_0	D_0	D_0	D_0	D_0	D_0	D_0	D_0	D_0	D_0	D_0	D_0	D_0	D_0	D_0	D_0	D_0			
SL 146 mm	R	B	B	B																					

EDITOR'S NOTE: For ease of use, Table 3 appears on page 85.

TABLE 4.—Distribution of intermuscular bones and ligaments in certain ostariophysan, esocoid, salmonoid, and osmeroid teleosts.

		Occ	ANA	1	2	3	4	5	6	7	8	9	10	11	12	13	14	15	16	17	18	19	20	21	22	23	24	25	26	27	28	29
Esox	En		L_1	F_1	F_1	F_1	F_1	B_1	B_1	$_vB_1$	$_vB_1$	$_vB_1$	$_vB_1$	$_vB_2$	$_vB_2$	$_vB_2$	$_vB_2$	$_vB_2$	$_vB_2$	$_vB_2$	$_vB_2$	$_vB_2$	$_vB_2$	$_vB_2$	$_vB_2$	$_vB_2$	$_vB_2$	$_vB_2$	$_vB_1$	$_vB_1$	$_vB_1$	$_vB_1$
lucius	Ec	L		L_2	L_4	L_4	L_4	L_4	L_4	L_4	L_4	L_4	L_4	L_4	L_4	L_4	L_4	L_4	L_4	L_4	L_4	L_4	L_4	L_4	L_4	L_4	L_4	L_4	L_4	L_4	L_4	L_4
ROM 598CS	Ep																												L_4	L_4	L_4	L_4
SL ca 70 mm	R			S	F_3	F_3	F_3	$\underline{B}$	$\underline{B}$	$\underline{B}$	$\underline{B}$	$\underline{B}$	$\underline{B}$	$\underline{B}$	$\underline{B}$	$\underline{B}$	$\underline{B}$	$\underline{B}$	$\underline{B}$	$\underline{B}$	$\underline{B}$	$\underline{B}$	$\underline{B}$	$\underline{B}$	$\underline{B}$	$\underline{B}$	$\underline{B}$	$\underline{B}$	$\underline{B}$	$\underline{B}$	$\underline{B}$	$\underline{B}$
Esox	En		B_1	F_1	F_1	V_1	V_0	V_0	$_vB_0$	$_vB_1$	$_vB_1$	$_vB_2$	$_vB_2$	$_vB_2$	$_vB_2$	$_vB_2$	$_vB_2$	$_vB_2$	$_vB_2$	$_vB_2$	$_vB_2$	$_vB_2$	$_vB_2$	$_vB_2$	$_vB_2$	$_vB_2$	$_vB_2$	$_vB_2$	$_vB_1$	$_vB_1$	$_vB_1$	$_vB_1$
americanus	Ec	L		L_2	L_4	L_4	L_4	L_4	L_4	L_4	L_4	L_4	L_4	L_4	L_4	L_4	L_4	L_4	L_4	L_4	L_4	L_4	L_3	L_3	L_3	L_2	L_2	L_2	L_2	L_2	L_2	L_2
BMNH 82.11.60.16	Ep																					L_4	L_4	B_4	B_4	B_4	B_4	B_4	B_3	B_3	B_3	B_3
SL 61 mm	R			S	F_3	$\underline{B}$	$\underline{B}$	$\underline{B}$	$\underline{B}$	$\underline{B}$	$\underline{B}$	$\underline{B}$	$\underline{B}$	$\underline{B}$	$\underline{B}$	$\underline{B}$	$\underline{B}$	$\underline{B}$	$\underline{B}$	$\underline{B}$	$\underline{B}$	$\underline{B}$	$\underline{B}$	B	B	B	B	B	B	B	B	B
Umbra	En		-	L_1	B_1	B_1	B_1	B_1	B_1	B_1	B_1	B_1	B_1	B_1	B_1	B_1	B_1	B_1	B_1	B_1	B_1	B_1	B_1	B_1	$>B_1$	B_0	B_0	B_0	B_0	B_0	B_0	B_0
krameri	Ec	L		L_3	L_4	L_4	L_4	L_4												L_4	L_4	L_3	L_2	L_2	L_2	L_2	L_2	L_2	L_2	L_2		
USNM 205523	Ep																				L_4	L_4	B_3	B_3	B_5	$>B_3$	B_3	B_0	B_0	B_0	B_0	B_0
SL 83 mm	R			S	B	$\underline{B}$	$\underline{B}$	$\underline{B}$	$\underline{B}$	$\underline{B}$	$\underline{B}$	$\underline{B}$	$\underline{B}$	$\underline{B}$	$\underline{B}$	$\underline{B}$	$\underline{B}$	$\underline{B}$	$\underline{B}$	$\underline{B}$	$\underline{B}$	$\underline{B}$	B									
Novumbra	En	L	-	L_1	L_1	L_1	L_1	L_1	L_1	L_1	L_1	L_1	L_1	L_1	L_1	L_1	L_1	L_1	L_1	L_1	L_1	L_1	L_1	L_1	L_1	L_1	L_1	L_1	L_1	L_1		
hubbsi	Ec																															**PU9**
AMNH 45019	Ep																				L_3	L_5	L_5	L_5	L_5	L_5	L_5	L_5	L_5	L_5	L_5	
SL 37 mm	R			S	B	$\underline{B}$	$\underline{B}$	$\underline{B}$	$\underline{B}$	$\underline{B}$	$\underline{B}$	$\underline{B}$	$\underline{B}$	$\underline{B}$	$\underline{B}$	$\underline{B}$	$\underline{B}$	$\underline{B}$	$\underline{B}$	$\underline{B}$	$\underline{B}$											
Chanos	En	VVV	-	V_0	$_vB_1$	$_vB_1$	$_vB_1$	$_vB_1$	$_vB_1$	$_vB_1$	$_vB_1$	$_vB_1$	$_vB_1$	$_vB_1$	$_vB_1$	$_vB_1$	$_vB_1$	$_vB_1$	$_vB_1$	$_vB_1$	$_vB_1$	$_vB_1$	$_vB_1$	$_vB_1$	$_vB_1$	$_vB_1$	$_vB_1$	$_vB_1$	$_vB_1$	$_vB_1$	$_vB_1$	$_vB_1$
chanos	Ec	$>B>B$		$<B_2$	$<B_2$	$<B_4$	$<B_4$	$<B_4$	$<B_4$	$<B_4$	$<B_4$	$<B_4$	$<B_4$	$<B_4$	$<B_4$	$<B_4$	$<B_3$	$<B_3$	$<B_3$	B_3	B_2	B_2	B_2	B_2	B_2	B_2	B_2	B_2	L_2	L_2	L_2	L_2
BMNH uncat.	Ep																		D_4	D_4	D_4	D_4	$_DB_3$	$_DB_3$	$_DB_3$	$_DB_3$	$_DB_3$	$_DB_3$	$_DB_3$	$_DB_3$	$_DB_3$	$_DB_3$
SL 49 mm	R					$\underline{B}$	$\underline{B}$	$\underline{B}$	$\underline{B}$	$\underline{B}$	$\underline{B}$	$\underline{B}$	$\underline{B}$	$\underline{B}$	$\underline{B}$	$\underline{B}$	$\underline{B}$	$\underline{B}$	$\underline{B}$	$\underline{B}$	$\underline{B}$	$\underline{B}$	$\underline{B}$	$\underline{B}$	$\underline{B}$	$\underline{B}$	$\underline{B}$	$\underline{B}$	$\underline{B}$	$\underline{B}$	$\underline{B}$	$\underline{B}$
Gonorynchus	En	BBB	-	V_0	V_0	L_2V_0	B_2V_0	$_vB_1$	$_vB_1$	$_vB_1$	$_vB_1$	$_vB_1$	$_vB_1$	$_vB_1$	$_vB_1$	$_vB_1$	$_vB_1$	$_vB_1$	$_vB_1$	$_vB_1$	$_vB_1$	$_vB_1$	$_vB_1$	$_vB_1$	$_vB_1$	$_vB_1$	$_vB_1$	$_vB_1$	$_vB_1$	$_vB_1$	$_vB_1$	$_vB_1$
gonorynchus	Ec	BBB		$<B_3$	B_3		B_3	B_3	B_3	B_3	B_3	B_3	B_3	B_3	B_3	B_3	B_3	B_3	B_3	B_3	B_3	B_3	B_3	B_3	B_3	B_3	B_3	B_3	B_3	B_3	B_3	B_3
SIO 70-275	Ep													B_3	B_3	B_3	B_3	B_3	B_3	B_3	B_3	B_3	B_3	B_3	B_3	B_3	B_3	B_3	B_3	B_3	B_3	B_3
SL 146 mm	R	XX		X		$\underline{B}$	B	B	B	B	B	B	B	B	B	B	B	B	B	B	B	B	B	B	B	B	B	B	B	B	B	B
Parakneria	En		-		B_2	B_2	B_1	B_1	B_1	$_vB_1$	$_vB_1$	$_vB_1$	$_vB_1$	$_vB_1$	$_vB_1$	$_vB_1$	$_vB_1$	$_vB_1$	$_vB_1$	$_vB_1$	$_vB_1$	$_vB_1$	$_vB_1$	$_vB_1$	$_vB_1$	$_vB_1$	$_vB_1$	$_vB_1$	$_vB_1$	$_vB_1$	$_vB_1$	$_vB_1$
sp.	Ec	$<B$		B_2	$<B_2$	$<B_4$	$<B_4$	$<B_4$	$<B_4$	$<B_4$	$<B_4$	$<B_4$	$<B_4$	$<B_4$	$<B_4$	B_4	$<B_4$	$<B_4$	B_4	B_4	B_4	B_4	B_4	B_4	B_4	B_3	B_3	B_3	B_3	B_2	B_2	B_2
BMNH uncat.	Ep																			B_4	B_4	B_4	B_4	B_4	B_4	B_4	B_4	$_DB_3$	$_DB_3$	$_DB_5$	$_DB_5$	$_DB_5$
SL 45 mm	R					B	$\underline{B}$	$\underline{B}$	$\underline{B}$	$\underline{B}$	$\underline{B}$	$\underline{B}$	$\underline{B}$	$\underline{B}$	$\underline{B}$	$\underline{B}$	$\underline{B}$	$\underline{B}$	$\underline{B}$	$\underline{B}$	B	B	B	B	B	B	B	B				
Hemibarbus	En		-	L_1	B_1		B_1	B_1	$_vB_1$	$_vB_1$	$_vB_1$	$_vB_1$	$_vB_1$	$_vB_1$	$_vB_1$	$_vB_1$	$_vB_1$	$_vB_1$	$_vB_1$	$_vB_1$	$_vB_1$	$_vB_1$	$_vB_1$	$_vB_1$	$_vB_1$	$_vB_1$	$_vB_1$	$_vB_1$	$_vB_1$	$_vB_1$	$_vB_1$	$_vB_1$
barbus	Ec			L_1				L_4	L_4	L_4	L_4	L_4	L_4	L_4	L_4	L_4	L_4	L_3	L_3	L_3	L_3	L_3	L_3	L_3	L_2	L_2	L_2	L_2	L_2	L_2	L_2	L_2
USNM 45234	Ep																				B_4	B_4	B_4	$_DB_3$	$_DB_3$	$_DB_3$	$_DB_3$	$_DB_5$	$_DB_5$	$_DB_5$	$_DB_5$	$_DB_5$
SL 74 mm	R			S				$\underline{B}$	$\underline{B}$	$\underline{B}$	$\underline{B}$	$\underline{B}$	$\underline{B}$	$\underline{B}$	$\underline{B}$	$\underline{B}$	$\underline{B}$	$L\underline{B}$	$L\underline{B}$	$L\underline{B}$	$L\underline{B}$	$L\underline{B}$	$L\underline{B}$	$L\underline{B}$	$L\underline{B}$							
Hypentilium	En		-				$>V_0$	$>V_0$	$>V_0$	$<V_0$	$<V_0$	$<V_0$	$<V_0$	$<V_0$	$_vB_1$	$_vB_1$	$_vB_1$	$_vB_1$	$_vB_1$	$_vB_1$	$_vB_1$	$_vB_1$	$_vB_1$	$_vB_1$	$_vB_1$	$_vB_1$	$_vB_1$	$_vB_1$	$_vB_1$	$_vB_1$	$_vB_1$	$_vB_1$
nigricans	Ec			L_1				L_4	L_4	L_4	L_4	L_4	L_4	L_4	L_4	L_4	L_4	L_4	L_4	L_4	L_4	L_4	L_4	L_5	L_5	L_5	L_2	L_2	L_2	L_2	L_2	L_2
USNM 273085	Ep																					L_4	L_4	B_5	$<B_5$	$_DB_5$	$_DB_5$	$_DB_5$	$_DB_5$	$_DB_5$	$_DB_5$	$_DB_5$
SL 33 mm	R							$\underline{B}$	$\underline{B}$	$\underline{B}$	$\underline{B}$	$\underline{B}$	$\underline{B}$	$\underline{B}$	$\underline{B}$	$\underline{B}$	$\underline{B}$	$\underline{B}$	$\underline{B}$	$\underline{B}$	$\underline{B}$	$\underline{B}$	$\underline{B}$									
Acanthopsoides	En	VV	-	V_0	V_0	V_0	V_0	V_0	$_vB_1$	$_vB_1$	$_vB_1$	$_vB_1$	$_vB_1$	$_vB_1$	$_vB_1$	$_vB_1$	$_vB_1$	$_vB_1$	$_vB_1$	$_vB_1$	$_vB_1$	$_vB_1$	$_vB_1$	$_vB_1$	$_vB_1$	$_vB_1$	$_vB_1$	$_vB_1$	$_vB_1$	V_0	V_0	V_0
robertsi	Ec					L_3	L_3	L_4	L_4	L_4	L_4	L_4	L_4	L_4	L_3	L_3	L_3	L_3	L_3	L_2	L_2	L_2	L_2	L_2	L_2	L_2	L_2	L_2	L_2	L_2	L_2	L_2
USNM 230261	Ep																B_4	B_4	$_DB_4$	$_DB_4$	$_DB_3$	$_DB_3$	$_DB_3$	$_DB_3$	$_DB_3$	$_DB_3$	$_DB_3$	$_DB_3$	$_DB_3$	$_DB_3$	D_0	D_0
SL 26 mm	R	S		S				B	B	B	B	B	B	B	B	B	B	B	B	B												
Alestes	En	V	-	V_0	V_0	V_0	V_0	$_vB_1$	$_vB_1$	$_vB_1$	$_vB_1$	$_vB_1$	$_vB_1$	$_vB_1$	$_vB_1$	$_vB_1$	$_vB_1$	$_vB_1$	$_vB_1$	$_vB_1$	$_vB_1$	$_vB_1$	$_vB_1$	$_vB_1$	$_vB_1$	$_vB_1$	$_vB_1$	$_vB_1$	$_vB_1$	$_vB_1$	$_vB_1$	$_vB_1$
dentex	Ec				L_3	L_2		L_4	L_4	L_4	L_4	L_4	L_4	L_4	L_4	L_4	L_4	L_4	L_4	L_4	L_3	L_3	L_3	L_3	L_3	L_2	L_2	L_2	L_2	L_2		
BMNH 82.4.13.515	Ep																	D_0	D_0	D_0	$_DB_4$	$_DB_4$	$_DB_4$	$_DB_4$	$_DB_4$	$_DB_4$	$_DB_4$	$_DB_4$	$_DB_4$	$_DB_5$	$_DB_5$	$_DB_5$
SL 46 mm	R	S						$\underline{B}$	$\underline{B}$	$\underline{B}$	$\underline{B}$	$\underline{B}$	$\underline{B}$	$\underline{B}$	$\underline{B}$	$\underline{B}$	$\underline{B}$	$\underline{B}$	$\underline{B}$	$\underline{B}$	$\underline{B}$	$\underline{B}$	$\underline{B}$	$\underline{B}$	$\underline{B}$	$\underline{B}$	$\underline{B}$	$\underline{B}$	$\underline{B}$			
Salmo	En	L_1——	L_1	B_1	B_1	B_1	B_1	B_1	B_1	B_1	B_1	B_1	B_1	B_1	B_1	B_1	B_1	B_1	B_1	B_1	B_1	B_1	B_1	B_1	B_1	B_1	B_1	B_1	B_1	B_1	B_1	B_1
salar	Ec	L		$\underline{L}_1$	$\underline{L}_2$	$\underline{L}_4$	$\underline{L}_4$	$\underline{L}_4$	$\underline{L}_4$	$\underline{L}_4$	$\underline{L}_4$	$\underline{L}_4$	$\underline{L}_4$	$\underline{L}_4$	$\underline{L}_4$	$\underline{L}_4$	$\underline{L}_4$	$\underline{L}_4$	$\underline{L}_4$	$\underline{L}_4$	$\underline{L}_4$	$\underline{L}_4$	$\underline{L}_4$	$\underline{L}_4$	$\underline{L}_4$	$\underline{L}_4$	$\underline{L}_4$	$\underline{L}_3$	$\underline{L}_3$	$\underline{L}_3$	$\underline{L}_3$	$\underline{L}_2$
BMNH 81.9.22.9	Ep																											L_4	L_4	L_4	L_4	L_3
SL 57 mm	R			S		$\underline{B}$	$\underline{B}$	$\underline{B}$	$\underline{B}$	$\underline{B}$	$\underline{B}$	$\underline{B}$	$\underline{B}$	$\underline{B}$	$\underline{B}$	$\underline{B}$	$\underline{B}$	$\underline{B}$	$\underline{B}$	$\underline{B}$	$\underline{B}$	$\underline{B}$	$\underline{B}$	$\underline{B}$	$\underline{B}$	$\underline{B}$	$\underline{B}$	$\underline{B}$	$\underline{B}$	$\underline{B}$	$\underline{B}$	$\underline{B}$
Thymallus	En		F_1	F_1	B_1	B_1	B_1	B_1	B_1	B_1	B_1	B_1	B_1	B_1	B_1	B_1	B_1	B_1	B_1	B_1	B_1	B_1	B_1	B_1	B_1	B_1						
thymallus	Ec	L		L_1	B_2	B_4	F_4	B_4	B_4	B_4	B_4	L_4	L_4	L_4	L_4	L_4	L_4	L_4	L_4	L_4	L_4	L_4	L_4	L_4	L_4	L_3	L_3	L_3	L_2	L_2	L_2	L_2
BMNH 70.10.14.3	Ep																															
SL 70 mm	R			S		$\underline{B}$	$\underline{B}$	$\underline{B}$	$\underline{B}$	$\underline{B}$	$\underline{B}$	$\underline{B}$	$\underline{B}$	$\underline{B}$	$\underline{B}$	$\underline{B}$	$\underline{B}$	$\underline{B}$	$\underline{B}$	$\underline{B}$	$\underline{B}$	$\underline{B}$	$\underline{B}$	$\underline{B}$	$\underline{B}$	$\underline{B}$	$\underline{B}$	$\underline{B}$	$\underline{B}$	$\underline{B}$	$\underline{B}$	$\underline{B}$
Searsia	En	L	+	f_1	f_1	f_1	F_1	F_1	F_1	F_1	F_1	F_1	F_1	F_1	F_1	F_1	F_1	F_1	F_1	F_1	F_1	F_1	F_1	F_1	F_1	F_1	B_1	B_1	B_1	B_1		
koefoedi	Ec	L		L_2	L_2	L_2	L_2	L_2	L_2	L_2	L_2	L_2	L_2	L_2	L_2	L_2	L_2	L_2	L_2	L_2	L_2	L_2	L_2	L_2	L_2	L_2	L_2	L_2	L_2		**PU18**	
BMNH 57.11.4.5	Ep					B_3	B_3	B_3	B_3	B_3	B_3	B_3	B_3	B_3	B_3	B_3	B_3	B_3	B_3	B_3	B_3	B_3	B_3	B_3	B_3	B_3	B_3	B_5				
SL 127 mm	R			S	$\underline{B}$	$\underline{B}$	$\underline{B}$	$\underline{B}$	$\underline{B}$	$\underline{B}$	$\underline{B}$	$\underline{B}$	$\underline{B}$	$\underline{B}$	$\underline{B}$	$\underline{B}$	$\underline{B}$	$\underline{B}$	$\underline{B}$	$\underline{B}$	$\underline{B}$	$\underline{B}$	$\underline{B}$	$\underline{B}$	$\underline{B}$	$\underline{B}$	$\underline{B}$					
Argentina	En	LL	-	f_1	f_1	f_1	f_1	F_1	F_1	F_1	F_1	F_1	F_1	F_1	F_1	F_1	F_1	F_1	F_1	F_1	B_1	B_1	B_1	B_1	B_1	B_1	B_1	B_1	B_1	B_1	B_1	B_1
georgei	Ec	L		?	L_3	L_4	L_4	L_4	L_4	L_4	L_4	L_4	L_4	L_4	L_4	L_4	L_4	L_4	L_4	L_4	L_4	L_4	L_4	L_4	L_4	L_3	L_3	L_3	L_3	L_2	L_2	L_2
USNM 187834	Ep																				L_4	L_4	L_4	B_4	B_4	B_4	B_3	B_3	B_3	B_3	B_3	B_3
SL 130 mm	R			S		$\underline{B}$	$\underline{B}$	$\underline{B}$	$\underline{B}$	$\underline{B}$	$\underline{B}$	$\underline{B}$	$\underline{B}$	$\underline{B}$	$\underline{B}$	$\underline{B}$	$\underline{B}$	$\underline{B}$	$\underline{B}$	$\underline{B}$	B	B	B	B	B	B	B	B	B	B	B	B
Glossanodon	En	LL	-	f_1	f_1	f_1	F_1	F_1	F_1	F_1	F_1	F_1	F_1	F_1	F_1	F_1	F_1	F_1	F_1	F_1	F_1	F_1	F_1	B_1	B_1	B_1	B_1	B_1	B_1	B_1	B_1	B_1
struhsakeri	Ec			L_1	L_2	L_4	L_4	L_4	L_4	L_4	L_4	L_4	L_4	L_4	L_4	L_4	L_4	L_4	L_4	L_4	L_4	L_4	L_4	L_4	L_4	L_4	L_4	L_4	L_4	L_3	L_3	L_3
USNM 36418	Ep																L_4	L_4	L_4	L_4	L_4	L_4	L_4	L_4	L_4	L_4	L_4	L_4	L_4	B_4	L_4	L_4
SL 110 mm	R			S		$\underline{B}$	$\underline{B}$	$\underline{B}$	$\underline{B}$	$\underline{B}$	$\underline{B}$	$\underline{B}$	$\underline{B}$	$\underline{B}$	$\underline{B}$	$\underline{B}$	$\underline{B}$	$\underline{B}$	$\underline{B}$	$\underline{B}$	$\underline{B}$	$\underline{B}$	$\underline{B}$	$\underline{B}$	$\underline{B}$	$\underline{B}$	$\underline{B}$	$\underline{B}$	B	B	B	B

TABLE 4.—Continued.

	30	31	32	33	34	35	36	37	38	39	40	41	42	43
Esoxl	$_VB_1$ L_4 L_4 B	$_VB_1$ L_4 L_4 B	$_VB_2$ L_3 L_4 B	$_VB_2$ L_3 L_4 B	$_VB_1$ L_3 L_4 B	$_VB_1$ L_2 L_4 B	$_VB_1$ L_2 B_3 B	$_VB_1$ L_2 B_3 B	$_VB_1$ L_2 B_3 B	$_VB_1$ L_2 B_3 B	$_VB_1$ L_2 B_3 B	$_VB_1$ L_2 B_3 B	$_VB_1$ L_2 B_3 B	B_1 L_2 L_5 B
Esoxa	$_VB_1$ L_2 B_3 B	$_VB_1$ L_2 B_3 B	B_1 L_2 B_3	B_1 L_2 B_3	B_1 B_5	B_1 B_5	B_0 B_0	PU9						
Umbra	B_0	PU4												
Chanos	V_1 L_2 $_DB_3$ B	V_1 L_2 D0	V_1 L_2 D_0	V_0 L_2 D_0	V_0 L_2 D_0	V_0 L_2 D_0	V_0 L_2 D_0	V_0 L_2 D_0	V_0 L_2 D_0	V_0 D_0	V_0 D_0	V_0 D_0	PU2	
Gonor	$_VB_1$ B_3 B_3D_0 B	$_VB_1$ B_3 B_3D_0 B	$_VB_1$ B_3 B_3D_0 B	$_VB_1$ B_3 B_3D_0 B	$_VB_1$ B_3 B_3D_0 B	$_VB_1$ B_3 B_3D_0	B_1V_0 B_3 B_3D_0	B_1V_0 B_3 D_0	B_1V_0 B_3 D_0	B_1V_0 B_3 D_0	B_1V_0 B_3 D_0	L_1V_0 L_3 D_0	V_0 L_3 D_0	V_0 L_5 D_0
Parakn	$_VB_1$ B_2 $_DB_5$	$_VB_1$ $_DB_5$	$_VB_1$ $_DB_5$	V_0 $_DB_5$	V_0 D_0	V_0 D_0	V_0 D_0	V_0 D_0	V_0 D_0	V_0 D_0	PU3			
Hemibar	$_VB_1$ L_2 $_DB_5$	$_VB_1$ L_2 $_DB_5$	$_VB_1$ L_2 $_DB_5$	$_VB_1$ L_2 D_5	V_1 L_2 D_5	V_1 L_2 D_5	V_0 L_2 D_0	V_0 L_2 D_0	V_0 L_2 D_0	V_0 $<D_0$	$<V_0$ $<D_0$	$<V_0$ D_0	$<V_0$ D_0	V_0
Hypent	$_VB_1$ L_2 $_DB_5$	$_VB_1$ L_2 $_DB_5$	V_1 L_2 $_DB_5$	V_1 L_2 D_5	V_1 L_2 D_5	V_1 L_2 D_0	V_0 L_2 D_0	V_0 L_2 D_0	V_0 D_0	V_0 D_0	V_0 D_0	V_0 D_0	V_0	V_0
Acanth	V_0 D_0	V_0 D_0	V_0 D_0	V_0 D_0	V_0 D_0	V_0 D_0	V_0 D_0	PU2						
Aleste	$_VB_1$ $_DB_5$	$_VB_1$ D_0	V_0 D_0	V_0 D_0	V_0 D_0	V_0 D_0	V_0 D_0	V_0 D_0	V_0 D_0	V_0 D_0	V_0 D_0	V_0 D_0	V_0 D_0	V_0 D_0
Salmo	B_1 L_2 L_3 B	B_1 L_2 L_3 B	B_1 L_2 L_3 B	L_2 L_3 B	L_2 L_3 B	L_2 L_3 B	L_2 L_5	L_2 L_5	L_2 L_5	L_2 L_5	L_2 L_5	L_2 L_5	L_2 L_5	L_2 L_5
Thymal	L_2 B	L_2 B	L_2 B	L_2 L_4 B	L_2 L_4 B	L_2 L_2 B	L_2 L_3 B	L_2 L_5	L_2 L_5	L_2 L_5	L_2 L_5	L_2 L_5	L_2 L_5	L_2 L_5
Argentin	B_1 L_2 B_3 B	B_1 L_2 B_3 B	B_1 L_3 B_3	L_1 L_3 L_3	L_1 L_3 L_5	L_1 L_3 L_5	L_1 L_3 L_5	L_1 L_2 L_5	L_1 L_2 L_5	L_1 L_2 L_5	L_1 L_2 L_5	L_2	L_2	L_2
Glossano	B_1 L_3 B_0 B	B_1 L_3 B_0 B	B_1 L_3 B_0 B	L_1 L_3 B_0 B	L_1 L_3 L_3 B	L_1 L_3 L_3 B	L_1 L_3 L_3	L_1 L_2 L_3	L_1 L_2 L_3	L_1 L_2 L_3	L_1 L_2 L_3	L_1 L_2 L_3	L_1 L_2 L_3	L_2

	44	45	46	47	48	49	50	51	52	53	54	55	56	57	58
Esoxl	B_1 L_2 L_5	B_1 L_2 L_5	B_1 L_2 L_5	B_1 L_2 L_5	L_1 L_2 L_5	L_1 L_2	L_1 L_2	L_2	PU11						
Esoxa															
Umbra															
Chanos															
Gonor	V_0 L_2 D_0	V_0 L_2 D_0	V_0 L_2 D_0	V_0 L_2 D_0	L_2 D_0	L_2 D_0	L_2 D_0	D_0	D_0	PU3					
Parakn															
Hemibar	PU2														
Hypent	PU5														
Acanth															
Aleste	PU3														
Salmo	L_2 L_5	L_2 L_5	L_5	PU12											
Thymal	L_2 L_5	L_2 L_5	L_2 L_5	L_5	PU9										
Argentin	L_2	L_2	**L_2**	PU4											
Glossano	L_2	L_2	PU7												

TABLE 4.—Continued.

		Occ	ANA	1	2	3	4	5	6	7	8	9	10	11	12	13	14	15	16	17	18	19	20	21	22	23	24	25	26	27	28	29
Plecoglossus	En |		+	B_1	B_1	B_1	B_1	B_1	B_1	B_1	B_1	B_1	B_1	B_1	B_1	B_1	B_1	B_1	B_1	B_1	B_1	B_1	B_1	B_1	B_1	B_1	B_1	B_1	B_1	B_1	B_1	B_1
altivelis	Ec |	$\underline{LL}$		$\underline{L}_1$	$\underline{L}_2$	$\underline{L}_4$	$\underline{L}_4$	$\underline{L}_4$	$\underline{L}_4$	$\underline{L}_4$	$\underline{L}_4$	$\underline{L}_4$	$\underline{L}_4$	$\underline{L}_4$	$\underline{L}_4$	$\underline{L}_4$	$\underline{L}_4$	$\underline{L}_4$	$\underline{L}_4$	$\underline{L}_4$	$\underline{L}_4$	$\underline{L}_4$	$\underline{L}_4$	$\underline{L}_4$	$\underline{L}_4$	$\underline{L}_4$	$\underline{L}_4$	$\underline{L}_4$	L_4	L_4	L_4	$\underline{L}_3$
BMNH 84.12.6.16	Ep |																															
SL 51 mm	R |			S		$\underline{B}$	$\underline{B}$	$\underline{B}$	$\underline{B}$	$\underline{B}$	$\underline{B}$	$\underline{B}$	$\underline{B}$	$\underline{B}$	$\underline{B}$	$\underline{B}$	$\underline{B}$	$\underline{B}$	$\underline{B}$	$\underline{B}$	$\underline{B}$	$\underline{B}$	$\underline{B}$	$\underline{B}$	$\underline{B}$	$\underline{B}$	$\underline{B}$	$\underline{B}$	$\underline{B}$	$\underline{B}$	$\underline{B}$	$\underline{B}$
Osmerus	En |	LL	+	B_1	B_1	B_1	B_1	B_1	B_1	B_1	B_1	B_1	B_1	B_1	B_1	B_1	B_1	B_1	B_1	B_1	B_1	$<B_1$	B_1	$<B_1$	B_1	B_1	B_1	B_1	B_1	B_1	B_1	B_1
mordax	Ec |	LL		L_2	L_2	L_3	L_3	L_4	L_4	L_4	L_4	L_4	L_4	L_4	L_4	L_4	L_4	L_4	L_4	L_4	L_4	L_4	L_4	L_4	L_4	L_4	L_4	L_4	L_4	L_4	L_4	L_3
BMNH 84.11.29.11	Ep |																															
SL 85 mm	R |			S	B	$\underline{B}$	$\underline{B}$	$\underline{B}$	$\underline{B}$	$\underline{B}$	$\underline{B}$	$\underline{B}$	$\underline{B}$	$\underline{B}$	$\underline{B}$	$\underline{B}$	$\underline{B}$	$\underline{B}$	$\underline{B}$	$\underline{B}$	$\underline{B}$	$\underline{B}$	$\underline{B}$	$\underline{B}$	$\underline{B}$	$\underline{B}$	$\underline{B}$	$\underline{B}$	$\underline{B}$	$\underline{B}$	$\underline{B}$	$\underline{B}$
Hypomesus	En |		+	B_1	B_1	B_1	B_1	B_1	B_1	B_1	B_1	B_1	B_1	B_1	B_1	B_1	B_1	B_1	B_1	B_1	B_1	B_1	B_1	B_1	B_1	B_1	B_1	B_1	B_1	B_1	B_1	B_1
transpacificus	Ec |	$\underline{LL}$		$\underline{L}_2$	$\underline{L}_2$	$\underline{L}_4$	$\underline{L}_4$	$\underline{L}_4$	$\underline{L}_4$	$\underline{L}_4$	$\underline{L}_4$	$\underline{L}_4$	$\underline{L}_4$	$\underline{L}_4$	$\underline{L}_4$	$\underline{L}_4$	$\underline{L}_4$	$\underline{L}_4$	$\underline{L}_4$	$\underline{L}_4$	$\underline{L}_4$	$\underline{L}_4$	$\underline{L}_4$	$\underline{L}_4$	$\underline{L}_4$	$\underline{L}_4$	$\underline{L}_4$	$\underline{L}_4$	$\underline{L}_4$	L_3	L_3	L_3
BMMH 84.6.28.11	Ep |																															L_4
SL 70 mm	R |			S		$\underline{B}$	$\underline{B}$	$\underline{B}$	$\underline{B}$	$\underline{B}$	$\underline{B}$	$\underline{B}$	$\underline{B}$	$\underline{B}$	$\underline{B}$	$\underline{B}$	$\underline{B}$	$\underline{B}$	$\underline{B}$	$\underline{B}$	$\underline{B}$	$\underline{B}$	$\underline{B}$	$\underline{B}$	$\underline{B}$	$\underline{B}$	$\underline{B}$	$\underline{B}$	$\underline{B}$	$\underline{B}$	$\underline{B}$	$\underline{B}$
Salangichthys	En |		+				L_1	L_1	L_1	L_1	L_1	L_1	L_1	L_1	L_1	L_1	L_1	L_1	L_1	L_1	L_1	L_1	L_1	L_1	L_1	L_1	L_1	L_1	L_1	L_1	L_1	L_1
microdon	Ec |						L_3	L_3	L_3	L_3	L_3	L_3	L_3	L_3	L_3	L_3	L_3	L_3	L_3	L_3	L_3	L_3	L_3	L_3	L_3	L_3	L_3	L_3	L_3	L_3	L_3	L_3
BMNH uncat.	Ep |																															
SL 82 mm	R |			S					C	C	C	C	C	C	C	C	C	C	C	C	C	C	C	C	C	C							
Stokellia	En |		-			L_1	L_1	L_1	L_1	L_1	L_1	L_1	L_1	L_1	L_1	L_1	L_1	L_1	L_1	L_1	L_1	L_1	L_1	L_1	L_1	L_1	L_1	L_1	L_1	L_1	L_1	L_1
anisodon	Ec |	$\underline{L}$		$\underline{L}_2$	$\underline{L}_2$	$\underline{L}_2$	$\underline{L}_4$	$\underline{L}_4$	$\underline{L}_4$	$\underline{L}_4$	$\underline{L}_4$	$\underline{L}_4$	$\underline{L}_4$	$\underline{L}_4$	$\underline{L}_4$	$\underline{L}_4$	$\underline{L}_3$	$\underline{L}_3$	$\underline{L}_3$	$\underline{L}_3$	$\underline{L}_3$	$\underline{L}_3$	$\underline{L}_3$	$\underline{L}_3$	$\underline{L}_3$	$\underline{L}_3$	$\underline{L}_3$	$\underline{L}_2$	$\underline{L}_2$	$\underline{L}_2$	$\underline{L}_2$	$\underline{L}_2$
BMNH 84.4.30.11	Ep |																															
SL 71 mm	R |			S		$\underline{B}$	$\underline{B}$	$\underline{B}$	$\underline{B}$	$\underline{B}$	$\underline{B}$	$\underline{B}$	$\underline{B}$	$\underline{B}$	$\underline{B}$	$\underline{B}$	$\underline{B}$	$\underline{B}$	$\underline{B}$	$\underline{B}$	$\underline{B}$	$\underline{B}$	$\underline{B}$	$\underline{B}$	$\underline{B}$	$\underline{B}$	$\underline{B}$	$\underline{B}$	$\underline{B}$	$\underline{B}$	$\underline{B}$	$\underline{B}$
Lepidogalaxias	En |		-																													
salamandroides	Ec |			$<B_2$	$<B_2$	$<B_4$	$<B_4$	$<B_4$	$<B_4$	L_4	L_3	L_3	L_3	L_3	L_2	L_2					**PU27**											
AMNH 37674SW	Ep |																															
SL 38 mm	R |			S	B	B	B	B	B	B	B	B	B	B	B	B	B	B	B	B												
Galaxias	En |		-	B_1	B_1	B_1	B_1	B_1	B_1	B_1	B_1	B_1	B_1	B_1	B_1	B_1	B_1	B_1	B_1	B_1	B_1	B_1	B_1	B_1	B_1	B_1	B_1	B_1	B_1	B_1	B_1	B_1
zebratus	Ec |			$\underline{L}_1$	$\underline{L}_2$	$\underline{L}_4$	$\underline{L}_4$	$\underline{L}_4$	$\underline{L}_4$	$\underline{L}_4$	$\underline{L}_4$	$\underline{L}_4$	$\underline{L}_4$	$\underline{L}_4$	$\underline{L}_4$	$\underline{L}_4$	$\underline{L}_4$	$\underline{L}_4$	$\underline{L}_3$	$\underline{L}_3$	$\underline{L}_3$	$\underline{L}_3$	$\underline{L}_2$	$\underline{L}_2$	$\underline{L}_2$	L_2	L_2	L_2				
BMNH 75.12.29.544	Ep |																							B_3	B_3	B_3	B_3	B_5	B_5	B_5	B_5	B_5
SL 58 mm	R |	S______		S	$\underline{B}$	$\underline{B}$	$\underline{B}$	$\underline{B}$	$\underline{B}$	$\underline{B}$	$\underline{B}$	$\underline{B}$	$\underline{B}$	$\underline{B}$	$\underline{B}$	$\underline{B}$	$\underline{B}$	$\underline{B}$	$\underline{B}$	$\underline{B}$	$\underline{B}$	$\underline{B}$	$\underline{B}$	$\underline{B}$	$\underline{B}$	$\underline{B}$	$\underline{B}$					

TABLE 4.—Continued.

	30	31	32	33	34	35	36	37	38	39	40	41	42	43	44	45	46	47	48	49	50	51	52	53	54	55	56	57	58
Pleco	B_1	B_1																											
	L_3	L_3	L_2	L_2	L_2	L_2	L_2	L_2	L_2																				
	L_4	L_4	L_4	L_3	L_3	L_3	L_3	L_3	L_3	L_5	L_5	L_5	L_5	L_5	L_5	L_5	**PU17**												
	B	B	B	B	B	B	B	B	B																				
Osmerus	B_1	B_1	B_1	B_1	B_1	B_1	B_1	B_1	L_1	L_1	L_1	L_1	L_1	L_1	L_1	L_1	L_1	L_1	L_1	L_1									
	L_3	L_3	L_3	L_3	L_2	L_2	L_2	L_2	L_2	L_2	L_2	L_2	L_2	L_2	L_2	L_2	L_2	L_2	L_2	L_2	L_2	L_2	L_2	L_2	L_2	**PU9**			
					L_3	L_3	L_3	L_3	L_3	L_3	L_3	L_5	L_5	L_5	L_5	L_5	L_5	L_5	L_5	L_5									
	B	B	B	B	B	B	B	B	B	B	B																		
Hypom	B_1	B_1	B_1	L_1	L_1	L_1	L_1	L_1	L_1	L_1																			
	L_3	L_3	L_3								**PU16**																		
	L_4		L_4	L_3	L_3	L_5	L_5	L_5																					
	B	B	B	B	B																								
Salang	L_1	L_1	L_1	L_1	L_1	L_1	L_1	L_1	L_1	L_1	L_1	L_1	L_1	L_1	L_1	L_1	L_1	L_1											
	L_3	L_3	L_3	L_2	L_2	L_2	L_2	L_2	L_2	L_2	L_2	L_2	L_2	L_2	L_2	L_2	L_2	L_2	**PU14**										
														L_5	L_5	L_5	L_5												
Stokel	L_1	L_1	L_1	L_1	L_1	L_1	L_1	L_1	L_1																				
										PU20																			
	B	B	B	B	B	B	B	B	B																				
Galax	B_1	B_1																											
			PU13																										
	B_5	B_5																											

TABLE 5.—Distribution of intermuscular bones and ligaments in certain stomiiform, aulopiform, and myctophiform neoteleosts.

		Occ	ANA	1	2	3	4	5	6	7	8	9	10	11	12	13	14	15	16	17	18	19	20	21	22	23	24	25	26	27	28	29
Diplophos	En		+	F_1	F_1	F_1	F_1	F_1	B_1	B_1	B_1	B_1	B_1	B_1	B_1	B_1	B_1	B_1	B_1	B_1	B_1	B_1	B_1	B_1	B_1	B_1	B_1	B_1	B_1	B_1	B_1	B_1
taenia	Ec			L_2	L_2	L_3	L_4	L_4	L_4	L_4	L_4	L_4	L_4	L_4	L_4	L_4	L_4	L_4	L_4	L_4	L_4	L_4	L_4	L_4	L_4	L_4	L_4	L_3	L_3	L_3	L_3	L_3
MCZ 55469	Ep																														B_4	B_4
SL 111 mm	R			*S*			$\underline{B}$	$\underline{B}$	$\underline{B}$	$\underline{B}$	$\underline{B}$	$\underline{B}$	$\underline{B}$	$\underline{B}$	$\underline{B}$	$\underline{B}$	$\underline{B}$	$\underline{B}$	$\underline{B}$	$\underline{B}$	$\underline{B}$	$\underline{B}$	$\underline{B}$	$\underline{B}$	$\underline{B}$	$\underline{B}$	$\underline{B}$	$\underline{B}$	$\underline{B}$	$\underline{B}$	$\underline{B}$	$\underline{B}$
Maurolicus	En	L	-	F_1	F_1	F_1	F_1	F_1	F_1	F_1	F_1	F_1	B_1	B_1	B_1	B_1	B_1	B_1	B_1	B_1	B_1	B_1	B_1									
muelleri	Ec	L		L_2	L_2	$\underline{L}_4$	$\underline{L}_4$	$\underline{L}_4$	$\underline{L}_4$	$\underline{L}_4$	$\underline{L}_4$	L_4	L_4	L_2	L_2									**PU13**								
AMNH 37329SW	Ep														B_4	B_3	B_5	B_5	B_5	B_5	B_5	B_5	B_5									
SL 42 mm	R			S		$\underline{B}$	$\underline{B}$	$\underline{B}$	$\underline{B}$	$\underline{B}$	$\underline{B}$	$\underline{B}$	$\underline{B}$	$\underline{B}$	$\underline{B}$																	
Pollichthys	En	L	-	F_1	F_1	F_1	F_1	F_1	F_1	F_1	F_1	F_1	F_1	F_1	F_1	F_1	F_1	F_1	F_1													
mauli	Ec			L_2	L_3	L_4	L_4	L_4	L_4	L_4	L_4	L_4	L_4	L_4	L_4	L_4	L_4	L_3	L_3	L_2	L_2	L_2						**PU 18**				
BMNH 84.1.1.13	Ep																				L_5	L_5	L_5	L_5	L_5	L_5	L_5					
SL 50 mm	R			S		$\underline{B}$	$\underline{B}$	$\underline{B}$	$\underline{B}$	$\underline{B}$	$\underline{B}$	$\underline{B}$	$\underline{B}$	$\underline{B}$	$\underline{B}$	$\underline{B}$	$\underline{B}$	$\underline{B}$	$\underline{B}$													
Aulopus	En	L	+	b_1	b_1	b_1	B_1	B_1	B_1	B_1	B_1	B_1	B_1	B_1	B_1	B_1	B_1	B_1	B_1	B_1	B_1	B_1	B_1	B_1	B_1	B_1	B_1	B_1	B_1	B_1	B_1	B_1
filamentosus	Ec	L		L_1	L_2	L_2	L_2	L_2 L_E	L_E	L_E	L_E	L_E	L_E	L_E	L_E	L_E		L_E	L_E	L_E	L_E	L_E	L_3	L_3	L_3	L_2	L_2	L_2	L_2	L_2	L_2	L_2
USNM 292105	Ep				$\mathbf{B}_0$	B_3	B_3	B_3	B_3	B_3	B_3	B_3	B_3	B_3	B_3	B_3	B_3	B_3	B_3	B_3	B_3	B_3	B_3	B_3	B_3	B_3	B_3	B_3	B_3	B_3	B_3	B_3
SL 115 mm	R			S		B	B	B	B	B	B	B	B	B	B	B	B	B	B	B	B	B	B	B	B	B	B	B				
Pseudotrichonotus	En		-	b_1	b_1		B_1	B_1	B_1	B_1	B_1	B_1	B_1	B_1	B_1	B_1	B_1	B_1	B_1	B_1	B_1	B_1	B_1	B_1	B_1	B_1	B_1	B_1	B_1			
altivelis	Ec	L																	$_pL_3$	$_pL_3$	$_pL_3$	$_pL_3$	L_3	L_3	L_2	L_2	L_2	L_2	L_2	L_2	L_2	L_2
USNM 280366	Ep				$\mathbf{B}_2$	$\mathbf{B}_3$	$\mathbf{B}_3$	$\mathbf{B}_3$	$\mathbf{B}_3$	$\mathbf{B}_3$	$\mathbf{B}_3$	$\mathbf{B}_3$	$\mathbf{B}_3$	$\mathbf{B}_3$	$\mathbf{B}_3$	$\mathbf{B}_3$	$\mathbf{B}_3$	$\mathbf{B}_3$	$\mathbf{B}_3$	$\mathbf{B}_3$	$\mathbf{B}_3$	$\mathbf{B}_3$	B_3	B_3	B_3	B_3	B_3	B_5	B_5	B_5	B_5	B_5
SL 81 mm	R			S			B	B	B	B	B	B	B	B	B	B	B	B	B	B	B	B	B	B	B						
Chlorophthalmus	En		-	b_1	b_1	B_1	B_1	B_1	B_1	B_1	B_1	B_1	B_1	B_1	B_1	B_1	B_1	B_1	$_vB_1$	$_vB_1$	$_vB_1$	$_vB_1$	$_vB_1$	$_vB_1$	$_vB_1$	$_vB_1$	$_vB_1$	$_vB_1$	$_vB_1$	$_vB_1$	$_vB_1$	$_vB_1$
bicornis	Ec																L_2	L_2	L_2	L_2	L_2	L_2	L_2	L_2	L_2	L_2	L_2	L_2	L_2	L_2	L_2	L_2
BMNH 86.8.21.10	Ep				$\mathbf{B}_2$	$\mathbf{B}_2$	$\mathbf{B}_3$	$\mathbf{B}_3$	$\mathbf{B}_3$	$\mathbf{B}_3$	B_3	B_3	B_3	B_3	B_3	B_3	B_3	B_3	B_3	B_3	B_5	$_DB_5$	$_DB_5$	$_DB_5$	$_DB_5$	$_DB_5$	$_DB_5$	$_DB_5$	$_DB_5$	$_DB_5$	$_DB_5$	$_DB_5$
SL 65 mm	R			S		$\underline{B}$	$\underline{B}$	$\underline{B}$	$\underline{B}$	$\underline{B}$	$\underline{B}$	B	B	B	B	B	B	B	B	B	B	B										
Bathypterois	En		-	B_1	B_1	B_1	B_1	B_1	B_1	B_1	B_1	B_1	B_1	B_1	B_1	B_1	B_1	B_1	B_1	B_1	B_1	B_1	B_1	B_1	B_1	B_1	B_1	B_1	B_1	B_1	B_1	B_1
pectinatus	Ec			L_1	L_1	L_1	L_1	L_2	L_3	L_3	L_3	L_3	L_3	L_3	L_3	L_3	L_3	L_3	L_3	L_3	L_3	L_2	L_2	L_2	L_2	L_2	L_2	L_2	L_2	L_2	L_2	L_2
FMNH 88982	Ep				B_3	B_3	B_3	B_3	B_3	B_3	B_3	B_3	B_3	B_3	B_3	B_3	B_3	B_3	B_3	B_3	B_3	B_3	B_3	B_3	B_3	B_3	B_3	B_3	B_3	B_3	B_3	B_3
SL 109 mm	R			S	B	B	$\underline{B}$	$\underline{B}$	$\underline{B}$	$\underline{B}$	$\underline{B}$	$\underline{B}$	B	B	B	B	B	B	B	B	B	B	B	B	B	*B*	*B*					
Parasudis	En		-	B_1	B_1	B_1	B_1	B_1	B_1	B_1	B_1	B_1	B_1	$_vB_1$	$_vB_1$	$_vB_1$	$_vB_1$	$_vB_1$	$_vB_1$	$_vB_1$	$_vB_1$	$_vB_1$	$_vB_1$	$_vB_1$	$_vB_1$	$_vB_1$	$_vB_1$	$_vB_1$	$_vB_1$	$_vB_1$	$_vB_1$	$_vB_1$
truculentus	Ec	*B*		B_2	B_2	B_2	B_2	B_3	B_3	B_3	B_3	B_3	B_2	B_2	B_2	B_2	B_2	B_2	B_2	B_2	B_2	B_2	B_2	B_2	B_2	B_2	B_2	B_2	B_2	B_2	B_2	B_2
USNM 159407	Ep				B_2	B_2	B_3	B_3	B_3	B_3	B_3	B_3	B_3	B_3	B_3	B_3	B_3	B_3	B_3	B_3	$_DB_3$	$_DB_5$	$_DB_5$	$_DB_5$	$_DB_5$	$_DB_5$	$_DB_5$	$_DB_5$	$_DB_5$	$_DB_5$	$_DB_5$	$_DB_5$
SL 108 mm	R			S		$\underline{B}$	$\underline{B}$	$\underline{B}$	$\underline{B}$	$\underline{B}$	$\underline{B}$	$\underline{B}$	$\underline{B}$	$\underline{B}$	$\underline{B}$	$\underline{B}$	$\underline{B}$	$\underline{B}$	$\underline{B}$	B	B											
Scopelosaurus	En	L	-	B_1	B_1	B_1	B_2	B_2	B_3	B_3	B_3	B_3	B_3	B_3	B_2	B_2	B_2	B_2	B_1	B_1	B_1	B_1	B_1	B_1	B_1	B_1	B_1	B_1	B_1	B_1	B_1	B_1
hubbsi	Ec	L																					$_pL_3$	$_pL_2$	$_pL_2$	$_pL_2$	$_pL_2$	L_2	L_2	L_2	L_2	L_2
BMNH 83.5.17.8	Ep				$\mathbf{B}_3$	$\mathbf{B}_3$	$\mathbf{B}_3$	$\mathbf{B}_3$	$\mathbf{B}_3$	$\mathbf{B}_3$	$\mathbf{B}_3$	$\mathbf{B}_3$	$\mathbf{B}_3$	$\mathbf{B}_3$	$\mathbf{B}_3$	$\mathbf{B}_3$	$\mathbf{B}_3$	$\mathbf{B}_3$	$\mathbf{B}_3$	$\mathbf{B}_3$	$\mathbf{B}_3$	$\mathbf{B}_3$	$<\mathbf{B}_3$	$<\mathbf{B}_3$	$<\mathbf{B}_3$	$<\mathbf{B}_3$	$<\mathbf{B}_3$	B_3	B_3	B_3	B_3	B_5
SL 76 mm	R			S		$\underline{B}$	$\underline{B}$	$\underline{B}$	$\underline{B}$	$\underline{B}$	$\underline{B}$	$\underline{B}$	$\underline{B}$	$\underline{B}$	$\underline{B}$	$\underline{B}$	$\underline{B}$	$\underline{B}$	$\underline{B}$	$\underline{B}$	$\underline{B}$	$\underline{B}$	$\underline{B}$	B	B	B	B	B	B	B		
Ahliesaurus	En	LL	-	B_1	B_1	B_1	B_2	B_2	B_2	B_2	B_2	B_2	B_2	B_2	B_2	B_2	B_2	B_2	B_2	B_2	B_2	B_2	B_2	B_2	B_2	B_2	B_2	B_2	B_2	B_2	B_1	B_1
berryi	Ec																	$_pL_3$	$_pL_3$	$_pL_2$	$_pL_2$	$_pL_2$	L_2	L_2	L_2	L_2	L_2	L_2	L_2	L_2	L_2	L_2
USNM 240505	Ep				$\mathbf{B}_3$	$\mathbf{B}_3$	$\mathbf{B}_3$	$\mathbf{B}_3$	$\mathbf{B}_3$	$\mathbf{B}_3$	$\mathbf{B}_3$	$\mathbf{B}_3$	$\mathbf{B}_3$	$\mathbf{B}_3$	$\mathbf{B}_3$	$\mathbf{B}_3$	$\mathbf{B}_3$	$\mathbf{B}_3$	$\mathbf{B}_3$	$\mathbf{B}_3$	$\mathbf{B}_3$	$<\mathbf{B}_3$	B_3	B_3	B_3	B_3	B_3	B_3	B_5	B_5	B_5	L_5
SL 77 mm	R			S		$\underline{B}$	$\underline{B}$	$\underline{B}$	$\underline{B}$	$\underline{B}$	$\underline{B}$	$\underline{B}$	$\underline{B}$	$\underline{B}$	$\underline{B}$	B	B	B	B	B	B	B	B	B	B	B	B	B				
Scopelarchoides	En	L	-	B_1	B_1	B_1	B_2	B_2	B_2	B_2	B_2	B_2	B_2	B_2	B_2	B_2	B_2	B_2	B_2	B_1	B_1	B_1	B_1	B_1	B_1	B_1	B_1	B_1	B_1	B_1	L_1	L_1
signifer	Ec	L			$_pL_3$	$_pL_3$	$_pL_3$	$_pL_3$	$_pL_3$	$_pL_3$	$_pL_3$	$_pL_3$	$_pL_3$	L_3	L_3	L_3	L_3	L_3	L_3	L_2	L_2	L_2	L_2	L_2	L_2	L_2	L_2	L_2	L_2	L_2	L_2	L_2
USNM 274385	Ep			$\mathbf{B}_3$	$\mathbf{B}_3$	$\mathbf{B}_3$	$\mathbf{B}_3$	$\mathbf{B}_3$	$\mathbf{B}_3$	$\mathbf{B}_3$	$\mathbf{B}_3$	$\mathbf{B}_3$	$\mathbf{B}_3$	B_3	B_3	B_3	B_3	B_3	B_3	B_3	B_5	B_5	B_5	B_5	B_5	B_5	B_5	B_5	L_5	L_5	L_5	L_5
SL 76 mm	R			**S**		$\underline{B}$	B	B	B	B	B	B	B	B	L_3	L_3																
Coccorella	En	L	-	B_1	L_1	L_1	L_1	L_1	L_2	L_2	L_2	L_2	L_2	L_2	L_2	L_2	L_1	L_1	L_1	L_1	L_1	L_1	L_1	L_1	L_1	L_1	L_1	L_1	L_1	L_1	L_1	L_1
atlanticum	Ec								$_pL_3$	$_pL_3$	$_pL_3$	$_pL_3$	$_pL_3$	$_pL_3$	$_pL_3$	L_3	L_3	L_3	L_2	L_2	L_2	L_2	L_2	L_2	L_2	L_2	L_2	L_2	L_2			
USNM 235199	Ep			$\mathbf{B}_2$	$\mathbf{B}_3$	$\mathbf{B}_3$	$\mathbf{B}_3$	$\mathbf{B}_3$	$\mathbf{B}_3$	$\mathbf{B}_3$	$\mathbf{B}_3$	$\mathbf{B}_3$	$\mathbf{B}_3$	$\mathbf{B}_3$	$\mathbf{B}_3$	B_3	B_3	B_3	L_5	L_5	L_5	L_5	L_5	L_5	L_5	L_5	L_5	L_5	L_5	L_5		
SL 60 mm	R			**S**		$\underline{B}$	$\underline{B}$	$\underline{B}$	$\underline{B}$	$\underline{B}$	$\underline{B}$	$\underline{B}$	$\underline{B}$	B	B	B	B	B														
Paralepis	En	$<B$	-	F_1	B_1	B_0	B_0	B_0	B_0	B_0	B_0	B_0	B_0	B_0	B_0	B_0	B_0	B_0	B_0	B_0	B_0	B_0	B_0	B_0	B_0	B_0	B_0	B_0	B_0	B_0	B_0	B_0
brevirostris	Ec	$<B$																														
USNM 196109	Ep	B_0		$\mathbf{F}_2$	$\mathbf{F}_2$	B_3	B_3	B_3	B_3	B_0	B_0	B_0	B_0	B_0	B_0	B_0	B_0	B_0	B_0	B_0	B_0	B_0	B_0	B_0	B_0	B_0	B_0	B_0	B_0	B_0	B_0	B_0
SL 110 mm	R			F**S**	F**S**	B	B	B	B	B	B	B	B	B	B	B	B	B	B	B	B	B	B	B	B	B	B	B	B	B	B	B
Macroparalepis	En		-	F_1	B_1	B_1	B_1	L_1	L_1	L_1																						
sp.	Ec																															
FMNH 49988	Ep			$\mathbf{B}_2$	$\mathbf{B}_2$	$\mathbf{B}_3$	$\boldsymbol{B}_3$	$\mathbf{B}_0$																								
SL 197 mm	R			**S**	**S**				B	B	B	B	B	B	B	B	B	B	B	B	B	B	B	B	B	B	B	B	B	B	B	B
Alepisaurus	En		-	F_1	F_1	F_1	F_1	F_1	B_0	B_0	B_0	B_0	B_0	B_0	B_0	B_0	B_0	B_0	B_0	B_0	B_0	B_0	B_0	B_0	B_0	B_0	B_0	B_0	B_0	B_0	B_0	B_0
ferox	Ec					B_2	B_2	B_2	B_2	B_2	B_2	B_2	B_2	B_2	B_2	B_2	B_2	B_2	B_2	B_2	B_2	B_2	B_2	B_2	B_2	B_2	B_2	B_2	B_2	B_2	B_2	B_2
MCZ 60345	Ep			$\mathbf{F}_2$	$\mathbf{F}_2$	$\mathbf{B}_3$	$\mathbf{B}_3$	$\mathbf{B}_3$	$\mathbf{B}_3$	$\mathbf{B}_3$	$\mathbf{B}_3$	$\mathbf{B}_3$	$\mathbf{B}_3$	$\mathbf{B}_3$	$\mathbf{B}_3$	$\mathbf{B}_3$	B_3	B_3	B_3	B_3	B_2	B_2	B_2	B_2	B_2	B_2	B_2	B_2	B_2	B_2	L_2	L_2
SL 95 mm	R			F	F**S**	B**S**	B**S**	B	B	B	B	B	B	B	B	B	B	B	B	B	B	B	B	B								
Synodus	En	B	+	$<b_1$	b_1	B_1	B_1	B_1	B_1	B_1	B_1	B_1	B_1	B_1	B_1	B_1	B_1	B_1	B_1	B_1	B_1	B_1	B_1	B_1	B_1	B_1	B_1	B_1	B_1	B_1	B_1	B_1
variegatus	Ec	L																													$_pL_3$	$_pL_3$	$_pL_3$	$_pL_3$	$_pL_3$
USNM 315318	Ep				$\mathbf{B}_2$	$\mathbf{B}_2$	B_3	$\mathbf{B}_3$	$\mathbf{B}_3$	$\mathbf{B}_3$	$\mathbf{B}_3$	$\mathbf{B}_3$	$\mathbf{B}_3$	$\mathbf{B}_3$	$\mathbf{B}_3$	$\mathbf{B}_3$	$\mathbf{B}_3$	$\mathbf{B}_3$	$\mathbf{B}_3$	$\mathbf{B}_3$	$\mathbf{B}_3$	$\mathbf{B}_3$	$\mathbf{B}_3$	$\mathbf{B}_3$	$\mathbf{B}_3$	$\mathbf{B}_3$	$\mathbf{B}_3$	$\mathbf{B}_3$	$\mathbf{B}_3$	$\mathbf{B}_3$	$\mathbf{B}_3$	$\mathbf{B}_3$
SL 78 mm	R			S				B	B	B	B	B	B	B	B	B	B	B	B	B	B	B	B	B	B	B	B	B	B	B	B	B

TABLE 5.—Continued.

	30	31	32	33	34	35	36	37	38	39	40	41	42	43	44	45	46	47	48	49	50	51	52	53	54	55	56	57	58	59	60	61	62	63	64	65
Diploph	B_1	B_1	B_1	B_1	B_1	B_1	B_1	B_1	B_1	B_1	B_1	B_1	B_1	L_1	L_1	L_1	L_1	L_1	L_1	L_1	L_1	L_1	L_1	L_1	L_1	L_1	L_1									
	L_3	L_3	L_3	L_2	L_2	L_2	L_2	L_2	L_2	L_2	L_2	L_2	L_2	L_2	L_2	L_2	L_2	L_2	L_2	L_2	L_2	L_2	L_2	L_2												**PU36**
	B_4	B_4	B_4	B_4	B_4	B_4	B_4	B_4	B_4	B_4	B_4	B_5	B_5	B_5	B_5	B_5	B_5	B_5	B_5	L_5	L_5	L_5	L_5	L_5	L_5	L_5	L_5	L_5	L_5	L_5	L_5	L_5	L_5	L_5	L_5	
	B	B	B	B	B	B	B	B	B	B	B																									
Aulopus	B_1	B_1	B_1	B_1	B_1	B_1																														
	L_2	L_2	L_2	L_2	L_2	L_2																														
	B_3	B_3	B_3	B_3	B_3	B_5	B_5	**PU14**																												
Pseudotr			**PU17**																																	
	B_5	B_5																																		
Chloroph	${}_{V}B_1$	${}_{V}B_1$	${}_{V}B_1$	${}_{V}B_1$	${}_{V}B_1$	${}_{V}B_1$	V_0	V_0	V_0	V_0	V_0	V_0	V_0																							
	L_2	L_2	L_2	L_2	L_2	L_2								**PU2**																						
	${}_{D}B_5$	${}_{D}B_5$	${}_{D}B_5$	${}_{D}B_5$	D_0	D_0	D_0	D_0	D_0	D_0	D_0	D_0	${}_{<}D_0$																							
Bathypte	B_1	B_1	B_1	B_1	B_1	B_1	B_1	B_1	B_1	B_1	B_1	B_1	B_1	B_1	B_1	B_1	B_1	B_1																		
	L_2	L_2	L_2	L_2	L_2	L_2	L_2	L_2	L_2	L_2	L_2	L_2	L_2	L_2	L_2	L_2	L_2		**PU9**																	
	B_5	B_5	B_5	B_5	B_5	B_5	B_5	B_5	B_5	B_5	B_5	B_5	B_5	B_5	B_5	B_5	B_5	B_5																		
Parasud	${}_{V}B_1$	${}_{V}B_1$	${}_{V}L_1$	${}_{V}L_1$	V_0	V_0	V_0	V_0																												
	L_2	B_2	L_2	L_2	L_2	L_2	L_2	L_2	**PU2**																											
	${}_{D}B_5$	${}_{D}L_5$	${}_{D}L_5$	D_0	D_0	D_0																														
Scopelos	B_1	B_1	B_1	B_1	B_1	B_1	B_1	B_1	L_1	L_1	L_1	L_1	L_1	L_1																						
	L_2	L_2	L_2	L_2	L_2	L_2	L_2	L_2	L_2	L_2						**PU14**																				
	B_5	B_5	B_5	B_5	B_5	B_5	L_5	L_5	L_5	L_5	L_5	L_5	L_5	L_5	L_5																					
Ahlie	B_1	L_1	L_1	L_1	L_1	L_1	L_1	L_1																												
	L_2	L_2	L_2	L_2	L_2	L_2	L_2			**PU10**																										
	L_5	L_5	L_5	L_5	L_5	L_5	L_5	L_5	L_5																											
Scopelar	L_1	L_1	L_1	L_1	L_1	L_1																														
											PU9																									
	L_5	L_5	L_5	L_5	L_5	L_5	L_5	L_5	L_5	L_5																										
Coccor	L_1	L_1	L_1																																	
				PU16																																
Paralep	B_0	B_0	B_0	B_0	B_0	B_0	B_0	B_0																												
									PU29																											
	B_0	B_0	B_0	B_0																																
	B	B	B	B	B	B	B	B																												
Macropa																																				
																PU59																				
	B	B	B	B	B	B	B	B	B	B	B	B	B	B	B																					
Alepisau	B_0	B_0																																		
	B_2															**PU7**																				
	L_2	L_2	L_2	L_2	L_2	L_2	L_2	L_2	L_2	L_2	L_2	L_2	L_2	L_2	L_2																					
Synodus	B_1	B_1	B_1	B_1	B_1	B_1	B_1	B_1	B_1	B_1	B_1	B_1	B_1	B_1	B_1	B_1	L_1	L_1	L_1																	
	${}_{P}L_3$	${}_{P}L_3$	${}_{P}L_2$	${}_{P}L_2$	${}_{P}L_2$	L_3	L_2	L_2	L_2	L_2	L_2	L_2	L_2	L_2	L_2	L_2	L_2	L_2	L_2	**PU12**																
	B_3	B_3	B_3	B_3	B_3	B_3	B_3	B_3	B_3	B_3	B_3	B_3	B_3	B_3	L_3	L_3	L_3	L_3																		
	B	B	B	B	B	B	B	B	B	B	B	B	B	B																						

TABLE 5.—Continued.

		Occ	ANA	1	2	3	4	5	6	7	8	9	10	11	12	13	14	15	16	17	18	19	20	21	22	23	24	25	26	27	28	29
Trachinocephalus	En	B	+	b_1	B_1	B_1	B_1	B_1	B_1	B_1	B_1	B_1	B_1	B_1	B_1	B_1	B_1	B_1	B_1	B_1	B_1	B_1	B_1	B_1	B_1	B_1	B_1	B_1	B_1	B_1	B_1	B_1
myops	Ec																								$_pL_3$	$_pL_3$	$_pL_3$	L_3	L_3	L_3	L_2	L_2
Anton Bruun uncat.	Ep				B_3	B_3	B_3	B_3	B_3	B_3	B_3	B_3	B_3	B_3	B_3	B_3	B_3	B_3	B_3	B_3	B_3	B_3	B_3	B_3	B_3	B_3	B_3	B_3	B_3	B_3	B_3	B_3
SL 105 mm	R			S				B	B	B	B	B	B	B	B	B	B	B	B	B	B	B	B	B	B	B	B	B	B	B	B	B
Harpadon	En		+	B_1	F_1	F_1	F_1	F_1	F_1	F_1	F_1	F_1	B_1	B_1	B_1	B_1	B_1	B_1	B_1	B_1	B_1	B_1	B_1	B_1	B_1	B_1	B_1	B_1	B_1	B_1	B_1	B_1
nehereus	Ec									$_pL_2$	$_pL_2$	$_pL_2$	$_pL_2$	$_pL_2$	$_pL_2$	L_2	L_2	L_2	L_2	L_2	L_2	L_2	L_2	L_2	L_2	L_2	L_2	L_2	L_2	L_2	L_2	L_2
FMNH 179018	Ep				B_2	B_3	B_3	B_3	B_3	B_3	B_3	B_3	B_3	B_3	B_3	B_3	B_3	B_3	B_3	B_3	B_3	B_3	B_3	B_3	B_3	B_3	B_3	B_3	B_3	B_3	B_3	B_3
SL 98 mm	R			S	B	B	B	B	B	B	B	B	B	B	B	B	B	B	B	B	B	B	B	B	B	B	B	B	B	B	B	B
Saurida	En		+	$_<F_1$	F_1	F_1	F_1	F_1	F_1	F_1	F_1	F_1	F_1	F_1	B_1	B_1	B_1	B_1	B_1	B_1	B_1	B_1	B_1	B_1	B_1	B_1	B_1	B_1	B_1	B_1	B_1	B_1
normani	Ec																						$_pL_2$	$_pL_2$	$_pL_2$	$_pL_2$	$_pL_2$	$_pL_2$	$_pL_2$	L_2	L_2	L_2
USNM uncat.	Ep				B_3	B_3	B_3	B_3	B_3	B_3	B_3	B_3	B_3	B_3	B_3	B_3	B_3	B_3	B_3	B_3	B_3	B_3	B_3	B_3	B_3	B_3	B_3	B_3	B_3	B_3	B_3	B_3
SL 95 mm	R			S			B	B	B	B	B	B	B	B	B	B	B	B	B	B	B	B	B	B	B	B	B	B	B	B	B	B
Bathysaurus	En		-	F_1	F_1	F_1	F_1	F_1	F_2	F_2	F_2	F_2	F_2	F_2	F_2	F_2	F_2	F_2	F_2	F_2	F_2	F_2	F_2	F_2	F_2	F_2	F_2	F_2	F_2	F_2	F_2	F_2
mollis	Ec																															
VIMS 06107	Ep			B_3	B_3	B_3	B_3	B_3	B_3	B_3	B_3	B_3	B_3	B_3	B_3	B_3	B_3	B_3	B_3	B_3	B_3	B_3	B_3	B_3	B_3	B_3	B_3	B_3	B_3	B_3	B_3	B_3
SL 182 mm	R			S	B	$L\underline{B}_E$	$L\underline{B}_E$	$L\underline{B}_E$	$L\underline{B}_E$	$L\underline{B}_E$																						
Gigantura	En	$_<B$	-	$_>B_2$	$_>B_2$	B_2	B_2	$_>B_2$	$_>B_2$	$_>B_2$	$_>B_2$	$_>B_2$	$_>B_2$	$_>B_2$	$_>B_2$	$_>B_2$	$_>B_2$	$_>B_2$	$_>B_2$	$_>B_2$	$_>B_2$	$_>B_2$	$_>B_2$	$_>B_2$	$_>B_2$	$_>B_2$	$_>B_2$	$_>B_2$	$_>B_2$	$_>B_0$	$_>B_0$	PU2
chuni	Ec																															
AMNH 55345SW	Ep			B_2	$_>B_2$	B_2	B_2	B_2	$_>B_2$	$_<B_2$	$_>B_2$	B_2	$B_{>2}$	B_2	B_2	$_>B_2$	B_2	B_2	B_2	B_2	B_2	B_2	B_2	B_2	B_2	B_2	B_2	B_2	B_2	B_0	B_0	
SL 168 mm	R																															
Myctophum	En		-	B_1	B_1	B_1	B_1	B_1	B_1	B_1	B_1	B_1	B_1	B_1	B_1	B_1	B_1	B_1	B_1	$_vB_1$	$_vB_1$	$_vB_1$	$_vB_1$	$_vB_1$	$_vB_1$	$_vB_1$	$_vB_1$	$_vB_1$	$_vB_1$	$_vB_1$	$_vB_1$	$_vB_1$
obtusirostre	Ec	L		L_2	L_2	L_4	L_4	L_4	L_4	L_4	L_4	L_4	L_4	L_3	L_3	L_3	L_3	L_5	L_2	L_2	L_2	L_2	L_2	L_2	L_2	L_2	L_2	L_2	L_2	L_2	L_2	L_2
AMNH 29140SW	Ep															B_4	B_3	B_3	B_5	B_5	B_5	B_5	$_DB_5$	$_DB_5$	$_DB_5$	$_DB_5$	$_DB_5$	$_DB_5$	$_DB_5$	$_DB_5$	$_DB_5$	$_DB_5$
SL 40 mm	R			S		$\underline{B}$	$\underline{B}$	$\underline{B}$	$\underline{B}$	$\underline{B}$	$\underline{B}$	$\underline{B}$	$\underline{B}$	$\underline{B}$	$\underline{B}$	$\underline{B}$	$\underline{B}$	$\underline{B}$														
Neoscopelus	En		-	B_1	B_1	B_1	B_1	B_1	B_1	B_1	B_1	B_1	B_1	B_1	B_1	B_1	B_1	$_vB_1$	$_vB_1$	$_vB_1$	$_vB_1$	$_vB_1$	$_vB_1$	$_vB_1$	$_vB_1$	$_vB_1$	V_0	V_0	V_0	V_0	PU4	
macrolepidotus	Ec			L_1	L_1	L_1	L_3	L_4	L_4	L_4	L_4	L_4	L_3	L_3	L_3	L_3	L_5	L_5	L_5	L_2	L_2	L_2	L_2	L_2	L_2	L_2	L_2	L_2				
AMNH 49533SW	Ep														B_3	B_3	B_5	B_5	$_DB_5$	$_DB_5$	$_DB_5$	$_DB_5$	$_DB_5$	$_DB_5$	$_DB_5$	D_1	D_0	D_0	D_0	D_0		
SL 103 mm	R			S		$\underline{B}$	$\underline{B}$	$\underline{B}$	$\underline{B}$	$\underline{B}$	$\underline{B}$	$\underline{B}$	$\underline{B}$	$\underline{B}$	$\underline{B}$	$\underline{B}$																

TABLE 5.—Continued.

	30	31	32	33	34	35	36	37	38	39	40	41	42	43	44	45	46	47	48	49	50	51	52
Trachin	B_1	B_1	B_1	B_1	B_1	B_1	B_1	B_1	B_1	B_1	B_1	B_1	B_1	B_1	L_1	L_1	L_1	L_1	L_1	L_1	PU5		
	L_2	L_2	L_2	L_2	L_2	L_2	L_2	L_2	L_2	L_2	L_2	L_2	L_2	L_2	L_2	L_2							
	B_3	B_3	B_3	B_3	B_3	B_3	B_3	B_3	B_3	B_3	B_3	B_3	B_3	B_3	B_3	L_3							
	B	B	B	B	B	B	B	B	B														
Harpado	B_1	B_1	B_1	B_1	PU9																		
	L_2	L_2																					
	B_3	B_3	B_3	L_3																			
	B																						
Saurida	B_1	B_1	B_1	B_1	B_1	B_1	B_1	L_1	L_1	PU11													
	L_2	L_2																					
	B_3	B_3	B_3	B_3	B_3	B_3	B_3	L_3															
	B	B	B	B																			
Bathys	F_2	F_2	F_2	F_2	B_2	B_2	B_2	L_2	PU12														
	B_3	B_3	B_3	B_3	B_3	B_3	B_3	L_3															
Myctoph	V_0	V_0	V_0	V_0	PU2																		
	L_2	L_2	L_2																				
	D_0	D_0	D_0	D_0																			

TABLE 6.—Matrix for 26 characters of the intermusculars and vertebral column in 19 genera of aulopiforms, the stomiiform *Diplophos*, the myctophid *Myctophum*, and an "ancestor" (a row of zeros). Except for character 25, the data are taken from Table 5, with the addition of *Omosudis*, based on USNM 219982, SL 81 mm. Character 25, the median caudal cartilages, is taken from Fujita (1990) and study of specimens; it was added to reinforce the monophyly of Synodontoidea s.l. (Aulopidae, Pseudotrichonotidae, Synodontidae, Harpadontidae; Johnson et al., 1989), which is otherwise not well supported by our intermuscular characters.

Taxon	Character				
			11111	11111	222222
	01234	56789	01234	56789	012345
Ancestor	00000	00000	00000	00000	000000
Diplophos	00100	00000	00000	00000	000001
Myctophum	10003	00100	00000	00031	000000
Aulopus	00000	11000	00111	01000	020001
Pseudotrichonotus	10000	11002	10122	01000	021001
Chlorophthalmus	10003	11102	10112	01031	010000
Bathypterois	10000	00000	00000	01000	013000
Parasudis	10003	00111	00000	01031	010000
Scopelosaurus	10010	00002	10112	11000	010000
Ahliesaurus	10010	00002	10112	11000	010000
Scopelarchoides	10010	00002	10112	02000	010020
Coccorella	10010	00002	10112	02000	010020
Paralepis	11102	00032	21112	02120	024100
Macroparalepis	11100	00032	20112	02010	022000
Alepisaurus	12102	00022	11112	02100	024100
Synodus	00000	11002	10122	01000	122011
Trachinocephalus	00000	10002	10122	01000	122011
Harpadon	03200	00002	10112	01000	023011
Saurida	03200	10002	10112	01000	021011
Gigantura	14031	00032	20111	02010	035010
Bathysaurus	10320	00032	20112	02000	013010
Omosudis	12102	00022	11112	02120	024100

Characters (* = run unordered):

0. ANA (accessory neural arch) present (0) or absent (1)
1*. Baudelot's ligament originates on V1 (0), or on V1 and V2 (1), or on V2-4 (2), or is ossified (3), or absent (4)
2*. Epineurals all independent (0), or the first five or fewer fused with the neural arch (1), or the first six or more fused with the neural arch (2), or the majority fused with the centrum (3)
3*. Epineurals all originate on the neural arch (0), or those on ~V5–15 originate on the centrum or parapophysis (1), or all beyond V5 originate on centrum (2), or all originate on centrum (3)
4*. Epineurals all attached to the axial skeleton (0), or a few unattached epineurals posteriorly (1), or the majority unattached (2), or the unattached epineurals represent only the anteroventral fork and posterior body of bone (notation V_0 in Table 5) (3)
5. First epineural "descended" (notation b_1 in Table 5) (1) or none descended (0)
6. First two or three epineurals "descended" (1) or one or none descended (0)
7. Some epineurals forked proximally (1) or none forked (0) (autapomorphous condition in *Gigantura* coded (0)
8*. All epicentrals in ligament (0), or all in bone from V1 (1), or all in bone from V3 (2), or epicentrals absent (3)
9. Epicentral on V1 in ligament (0), or in bone (1), or absent (2)
10. Epicentral series begins anteriorly at V1 (0), or more posteriorly (1), or absent (2)
11. Epipleurals on V1 and V2 fused with centrum (1) or free/absent (0)
12. Anterior epipleurals partially in (1) or entirely beneath (0) the horizontal septum
13*. Transition between epipleurals in and beneath the horizontal septum absent (none are in the septum) (0), or gradual (1), or abrupt (2)
14. Number of epipleurals with distal end in horizontal septum zero (0), one (1), or more (2)
15. One or more epipleurals forked distally at the transition between epipleurals in and beneath the horizontal septum (1), or no forking/no epipleurals in horizontal septum (0)
16. Epipleural series confined to the middle region of the body (0), or extends anteriorly to V2 (1), or to V1 (2)
17. Epipleural fused with V1 and V2 (1) or unfused (0)
18*. All epipleurals attached to the axial skeleton (0), or a few unattached epipleurals posteriorly (1), or the majority unattached (2), or the unattached epineurals represent only the anterodorsal fork and posterior body of the bone (notation D_0 in Table 5) (3)
19. Some epipleurals forked proximally (1), or all either unforked or forked only distally (0) (autapomorphous condition in *Gigantura* coded 0)
20. A brushlike posterodorsal outgrowth of the first neural arch present (1) or absent (0)
21*. All ribs ossified in cartilage (0), or some ossified in membrane bone (1), or all in membrane bone (2), or ribs absent (3)
22*. First rib originates on V3 (0), or V4 (1), or V5 (2), or V2 (3), or V1 (4), or ribs absent (5)
23. Ribs on V1 and V2 fused to centrum (1), or absent/free (0)
24*. Ratio of number of abdominal vertebrae to number of caudal vertebrae 0.25-1.75 (0), <0.25 (1), >1.75 (2)
25. Median caudal cartilages (CMC of Kafuku, 1990) present (0) or absent (1)

TABLE 7.—Distribution of intermuscular bones and ligaments in certain lower acanthomorph neoteleosts.

		Occ	ANA	1	2	3	4	5	6	7	8	9	10	11	12	13	14	15	16	17	18	19	20	21	22	23	24	25	26	//	30
Metavelifer	En		-	b_1	B_1	B_1	B_1	B_1	B_1																						
multiradiatus	Ec					L_4	L_4	L_4	L_4	L_3	L_3	L_3	L_3							PU18											
BPBM 23953	Ep																														
SL ca. 75 mm	R	S_____		S		$\underline{B}$	$\underline{B}$	$\underline{B}$	$\underline{B}$	$\underline{B}$	$\underline{B}$	$\underline{B}$	$\underline{B}$	$\underline{B}$	$\underline{B}$	$\underline{B}$	$\underline{B}$	$\underline{B}$	B												
Velifer	En		-	b_1	B_1	B_1	B_1	B_1	B_1	B_1	$\boldsymbol{B}_1$	L_1	L_1	L_1	L_1	L_1	L_1	L_1	L_1	L_1	L_1	L_1	L_1	L_1							
hypselopterus	Ec			L_2	L_2	L_4	L_4	L_4	L_4	L_4	L_4	L_4	L_4	L_4	L_4	L_3	L_3	L_3	L_2	L_2	L_2	L_2	L_2	L_2	L_2	L_2	L_2	L_2	PU7		
RUSI 13821	Ep																														
SL 133 mm	R	S		S		$\underline{B}$	$\underline{B}$	$\underline{B}$	$\underline{B}$	$\underline{B}$	$\underline{B}$	$\underline{B}$	$\underline{B}$	$\underline{B}$	$\underline{B}$	$\underline{B}$	B	B	$\boldsymbol{B}_5$	L_5	L_5	L_5	L_5	L_5	L_5	L_5	L_5				
Lampris	En		-	b_1	B_1	B_4	B_4	B_4	B_4	?>																					
guttatus	Ec					L_N	L_N	L_N	L_4	L_4	L_4	L_4	L_4	L_4	?>							PU25									
SIO 82-70	Ep																														
SL 167 mm	R			S		$\underline{B}$	$\underline{B}$	$\underline{B}$	$\underline{B}$	$\underline{B}$	$\underline{B}$	$\underline{B}$	$\underline{B}$	$\underline{B}$	$\underline{B}$	$\underline{B}$	$\underline{B}$	$\underline{B}$	$\underline{B}$	$\underline{B}$	$\underline{B}$										
Polymixia	En		-	B_1	B_1	B_3	B_3	B_3	B_3	B_3	B_3	B_2	B_2	B_1	B_1	B_1	B_1	B_1	$\boldsymbol{B}_1$	$\boldsymbol{B}_1$	$\boldsymbol{B}_1$	$\boldsymbol{B}_1$	$\boldsymbol{B}_1$	L_1	L_1	L_1	L_1				
lowei	Ec				L_1	$\underline{L}_4$	$\underline{L}_4$	$\underline{L}_4$	$\underline{L}_4$	$\underline{L}_4$	$\underline{L}_4$	$\underline{L}_3$	$\underline{L}_3$	$\underline{L}_3$	$\underline{L}_3$	$\underline{L}_5$	$\underline{L}_5$	$\underline{L}_5$	$\underline{L}_5$	$\underline{L}_5$	L_5	L_2	L_2	L_2	L_2	L_2	L_2	PU4			
BMNH 87.12.7.1	Ep										L_4	L_4	$\boldsymbol{B}_4$	B_4	B_4	B_5	B_5	B_5	B_5	B_5	$\boldsymbol{B}_5$	$\boldsymbol{B}_5$	L_5	L_5	L_5						
SL 128 mm	R			S		$\underline{B}$	$\underline{B}$	$\underline{B}$	$\underline{B}$	$\underline{B}$	$\underline{B}$	$\underline{B}$	$\underline{B}$	$\underline{B}$	$\underline{B}$	$\underline{B}$															
Percopsis	En		-	$\boldsymbol{B}_1$	$\boldsymbol{B}_1$	B_4	$\boldsymbol{B}_4$	$\boldsymbol{B}_4$																							
guttatus	Ec																			PU19											
BMNH 92.12.30.371	Ep																														
SL 40 mm	R			S		$\underline{B}$	$\underline{B}$	$\underline{B}$	$\underline{B}$	$\underline{B}$	$\underline{B}$	$\underline{B}$	$\underline{B}$	$\underline{B}$	$\underline{B}$	$\underline{B}$	$\underline{B}$	$\underline{B}$	$\underline{B}$												
Aphredoderus	En		-	B_1	B_2	B_4	B_4	B_4	B_4	B_4	$\boldsymbol{B}_4$																				
sayanus	Ec																PU16														
USNM 217374	Ep																														
SL 37 mm	R			S		$\underline{B}$	$\underline{B}$	$\underline{B}$	$\underline{B}$	$\underline{B}$	$\underline{B}$	$\underline{B}$	$\underline{B}$	$\underline{B}$	$\underline{B}$	$\underline{B}$															
Raniceps	En		-	B_0		B_4	B_4	B_4	B_4	B_4	B_4	B_3	B_3	$\boldsymbol{B}_3$	$\boldsymbol{B}_2$	$\boldsymbol{B}_2$	$\boldsymbol{B}_2$	L_2	L_2	L_2											
raninus	Ec																				ca PU28										
BMNH 93.7.6.2.	Ep																														
SL ca. 75 mm	R			S		$\underline{B}$	$\underline{B}$	$\underline{B}$	$\underline{B}$	$\underline{B}$	$\underline{B}$	$\underline{B}$																			
Gaidropsarus	En		-			B_4	B_4	B_4	B_4	B_4	B_4	B_4	B_4	B_3	B_3	B_3	B_3	B_3	B_5	B_5	B_5	B_2	B_2	B_2	B_2	$\boldsymbol{B}_2$	$\boldsymbol{B}_2$	$\boldsymbol{B}_2$	$\boldsymbol{B}_2$	//	
mediterraneus	Ec																														PU16
BMNH uncat.	Ep																														
SL 130 mm	R			S		$\underline{B}$	$\underline{B}$	$\underline{B}$	$\underline{B}$	$\underline{B}$	B	B	B	B	B	B	B														
Stephanoberyx	En		-																												
monae	Ec												PU21																		
USNM 208284	Ep																														
SL 88 mm	R			S	$\underline{B}$	$\underline{B}$	$\underline{B}$	$\underline{B}$	$\underline{B}$	$\underline{B}$	$\underline{B}$	$\underline{B}$																			
Gibberichthys	En		-	B_1	B_1	B_2	B_2	B_2	B_3	B_3	B_3																				
pumilus	Ec																PU16														
USNM 207512	Ep																														
SL 56 mm	R			S		$\underline{B}$	$\underline{B}$	$\underline{B}$	$\underline{B}$	$\underline{B}$	$\underline{B}$	$\underline{B}$	$\underline{B}$	$\underline{B}$	$\underline{B}$	$\underline{B}$															
Poromitra	En		-	B_1	B_2	B_3	B_3																								
oscitans	Ec										PU18																				
USNM 215625	Ep																														
SL 55 mm	R			S		$\underline{B}$	$\underline{B}$	$\underline{B}$	$\underline{B}$	$\underline{B}$																					
Hispidoberyx	En		-	B_1	B_1																										
ambiguus	Ec																	PU20													
AMS I.28080-001	Ep																														
SL 173 mm	R			S		$\underline{B}$	$\underline{B}$	$\underline{B}$	$\underline{B}$	$\underline{B}$	$\underline{B}$	$\underline{B}$	$\underline{B}$	$\underline{B}$	$\underline{B}$	$\underline{B}$	$\underline{B}$														
Rondeletia	En		-	$_{c}B_1$	$_{c}B_1$	$_{c}B_2$	$_{c}B_2$	$_{c}B_2$	$_{c}B_3$	$_{c}B_3$	$_{c}B_3$	$\boldsymbol{B}_3$	$\boldsymbol{B}_3$	L_5	L_5	L_5	L_5	L_2	L_2												
loricata	Ec																			PU9											
AMS I.20314-011	Ep																														
SL 38 mm	R	S_____		S		B	$\underline{B}$	$\underline{B}$	$\underline{B}$	$\underline{B}$	$\underline{B}$	$\underline{B}$	$\underline{B}$																		
Barbourisia	En	L	-	B_1	B_1	B_4	B_4	B_4																							
rufa	Ec				B_2																PU24										
AMS I.18824-001	Ep																														
SL 102 mm	R			S		$\underline{B}$	$\underline{B}$	$\underline{B}$	$\underline{B}$	$\underline{B}$	$\underline{B}$	$\underline{B}$	$\underline{B}$	$\underline{B}$	$\underline{B}$	$\underline{B}$	$\underline{B}$	$\underline{B}$	$\underline{B}$	B											
Diretmus	En		-	B_1																											
argenteus	Ec									L_5	L_5	L_5	L_5	L_5	L_5	L_5	L_5	L_5	L_5	L_2	L_2	PU11									
VIMS uncat.	Ep																														
SL 40 mm	R	S				$\underline{B}$	$\underline{B}$	$\underline{B}$	$\underline{B}$	$\underline{B}$	$\underline{B}$	$\underline{B}$	$\underline{B}$	$\underline{B}$	$\underline{B}$	B															
Anoplogaster	En		-																												
cornuta	Ec																PU14														
VIMS uncat.	Ep																														
SL 90 mm	R					$L\underline{B}$	$L\underline{B}$	$L\underline{B}$	$L\underline{B}$	$L\underline{B}$	$L\underline{B}$	$L\underline{B}$	$L\underline{B}$	$L\underline{B}$	$L\underline{B}$	$L\underline{B}$															

TABLE 7.—Continued.

		Occ	ANA	1	2	3	4	5	6	7	8	9	10	11	12	13	14	15	16	17	18	19	20	21	22	23	24	25	26	27	28
Monocentris	En		-								$\mathbf{B}_5$	$\mathbf{B}_5$	$\mathbf{B}_5$	$\mathbf{B}_5$	$\mathbf{L}_5$	$\mathbf{L}_5$															
japonicus	Ec													L_5	L_5	L_5	L_5	L_5	L_5	L_5	L_2	L_2	L_2	L_2	L_2	L_2	**PU3**				
USNM 307587	Ep																														
SL 53 mm	R							L_2	B_3	B_3	B_5	B_5	$\underline{\mathrm{B}}_5$	B_5	B_5																
Hoplostethus	En		-	$\mathbf{B}_1$	$\mathbf{B}_1$																										
sp.	Ec											L_5	L_5	L_5	L_5	L_2															
USNM 269460	Ep																														
SL 80 mm	R	S				$\underline{\mathrm{B}}$	$\underline{\mathrm{B}}$	$\underline{\mathrm{B}}$	$\underline{\mathrm{B}}$	$\underline{\mathrm{B}}$	$\underline{\mathrm{B}}$	$\underline{\mathrm{B}}$	$\underline{\mathrm{B}}$	B																	
Trachichthys	En		-	$\mathbf{B}_1$	$\mathbf{B}_1$					L_1	L_1	L_1	L_1	L_1	L_1	L_1															
australis	Ec									L_4	L_5	L_5	L_5	L_5	L_5	L_5	L_5	L_3	L_2	L_2	L_2	**PU7**									
AMS I.16861-004	Ep																														
SL 103 mm	R	S				$\underline{\mathrm{B}}$	$\underline{\mathrm{B}}$	$\underline{\mathrm{B}}$	B	B	$\underline{\mathrm{B}}$	$\underline{\mathrm{B}}$	B																		
Photoblepharon	En		-	$\mathbf{B}_1$	$\boldsymbol{B}_1$				$\boldsymbol{L}_4$	$\boldsymbol{L}_4$	$\boldsymbol{B}_3$	B_3	B_3	B_3	B_3	L_2	L_1	L_1													
palbebratus	Ec												L_3	L_3	L_3	L_3	L_5	L_5	L_5	L_5	L_5	L_5	L_2	L_2	L_2	L_2	L_2	**PU5**			
SIO 74-58	Ep																														
SL 54 mm	R	S				$\underline{\mathrm{B}}$	$\underline{\mathrm{B}}$	$\underline{\mathrm{B}}$	$\underline{\mathrm{B}}$	$\underline{\mathrm{B}}$	$\underline{\mathrm{B}}$	$\underline{\mathrm{B}}$	$\underline{\mathrm{B}}$	$\underline{\mathrm{B}}$	$\underline{\mathrm{B}}$	B															
Anomalops	En		-	$\mathbf{B}_1$	$\mathbf{B}_1$	$\mathbf{B}_4$	$\mathbf{B}_4$	$\mathbf{B}_3$	$\mathbf{B}_3$	B_3	B_3	B_3	B_3	B_2	L_2	L_1	L_1	L_1													
katoptron	Ec													L_4	L_4	L_3	L_3	L_5	L_5	L_5	L_5	L_2	L_2	L_2	L_2	L_2	L_2	L_2	L_2	**PU3**	
CAS 55163	Ep																														
SL 81 mm	R	S				$\underline{\mathrm{B}}$	$\underline{\mathrm{B}}$	$\underline{\mathrm{B}}$	$\underline{\mathrm{B}}$	$\underline{\mathrm{B}}$	$\underline{\mathrm{B}}$	$\underline{\mathrm{B}}$	$\underline{\mathrm{B}}$	$\underline{\mathrm{B}}$	$\underline{\mathrm{B}}$	B															
Centroberyx	En		-	$\mathbf{B}_1$	$\mathbf{B}_1$	$\mathbf{B}_4$	$\mathbf{B}_4$	$\mathbf{B}_4$	$\mathbf{B}_3$	B_2	B_2	B_2	B_2	L_2	L_1	L_1	L_1														
affinis	Ec											L_3	L_3	L_5	L_5	L_5	L_2	L_2	L_2	L_2	L_2	L_2	**PU5**								
USNM 176984	Ep																														
SL 110 mm	R	S				$\underline{\mathrm{B}}$	$\underline{\mathrm{B}}$	$\underline{\mathrm{B}}$	$\underline{\mathrm{B}}$	$\underline{\mathrm{B}}$	$\underline{\mathrm{B}}$	$\underline{\mathrm{B}}$	B																		
Holocentrus	En		-	$_{c}\mathbf{B}_1$	$_{c}\mathbf{B}_1$	$_{c}\mathbf{B}_4$	$_{c}\boldsymbol{B}_4$	$\boldsymbol{B}_4$	$\mathbf{B}_4$	$\mathbf{B}_4$	B_4	B_3	B_2	B_2	B_1	B_1	B_1	B_1													
spinifer	Ec											L_4	L_3	L_3	L_5	L_5	L_5	L_5	L_5	L_5	L_2	L_2	L_2	L_2	L_2	L_2	**PU4**				
BMNH 60.3.15.169	Ep														L_5	L_5															
SL 74 mm	R	S				$\underline{\mathrm{B}}$	$\underline{\mathrm{B}}$	$\underline{\mathrm{B}}$	$\underline{\mathrm{B}}$	$\underline{\mathrm{B}}$	$\underline{\mathrm{B}}$	$\underline{\mathrm{B}}$	$\underline{\mathrm{B}}$	B																	
Holocentrus	En		-	$\mathbf{B}_1$	$\mathbf{B}_1$	$\mathbf{B}_4$	$\mathbf{B}_4$	$\mathbf{B}_4$	$\mathbf{B}_3$	$\mathbf{B}_2$	B_1	B_1	B_1	B_1	B_1																
vexillaris	Ec											L_3	L_3	L_3	L_5	L_5	L_5	L_5	L_2	L_2	L_2	L_2	L_2	L_2	**PU6**						
USNM 269553	Ep														L_5	L_5															
SL 52 mm	E	S				$\underline{\mathrm{B}}$	$\underline{\mathrm{B}}$	$\underline{\mathrm{B}}$	$\underline{\mathrm{B}}$	$\underline{\mathrm{B}}$	$\underline{\mathrm{B}}$	$\underline{\mathrm{B}}$	$\underline{\mathrm{B}}$	B																	
Xenolepidichthys	En		-		$\mathbf{B}_1$	$\mathbf{B}_1$	$\mathbf{B}_3$	$\mathbf{B}_3$	$\mathbf{B}_3$	$\mathbf{B}_5$	$\mathbf{L}_5$																				
dalgleishi	Ec														L_5	L_5	L_5	L_5	L_5	L_5	L_5	L_5	L_5	**PU17**							
USNM 320016	Ep																														
SL 64 mm	R	S					L	L	L	L	L	B																			
Parazen	En		-	$\mathbf{B}_0$	$\mathbf{B}_0$	$\mathbf{B}_0$	$\mathbf{B}_0$	$\mathbf{B}_4$	$\mathbf{B}_4$	$\mathbf{B}_0$	$\mathbf{B}_0$	$\mathbf{B}_0$	$\mathbf{B}_0$	$\mathbf{B}_0$	$\mathbf{B}_0$	$\mathbf{B}_0$															
pacificus	Ec																**PU12**														
CAS 38404	Ep																														
SL 96 mm	R	S						L	L	L	L	L	L	B	B	B															

TABLE 8.—Distribution of intermuscular bones and ligaments in certain percomorph teleosts.

		Occ	ANA	1	2	3	4	5	6	7	8	9	10	11	12	13	14	15	16	17	18	19	20	21	22	23	24	25	//	29	30	31
Elassoma	En |		-	B_P	B_P	B_4	B_4	B_4	B_4	B_4	B_4	B_4	B_4	B_4	B_5	B_5																
sp.	Ec |																PU15															
USNM 313100	Ep |																															
SL 25 mm	R |	S				B	B	B	B	B	B	B	B	B																		
Agonostomus	En |		-	B_P	B_P	B_4	B_4	B_4	B_4																							
monticola	Ec |												L_3	L_5	L_5	L_5	L_2	L_2	L_2	L_2	L_2	PU6										
USNM 73742	Ep |																															
SL 32 mm	R |	S				B	B	B	B	B	B	B	B																			
Liza	En |		-	B_P	B_4	B_4	L_4																									
vaigiensis	Ec |										L_4	L_4	L_3	L_5	L_2	L_2	L_2	L_2	L_2	L_2	L_2	PU6										
BMNH 74.5.25.3646	Ep |																															
SL 19 mm	R |	S			B	B	B	B	B	B	B	B	B																		
Bedotia	En |		-	B_P	B_P	B_4	B_4	B_4	B_4	B_4	B_4	B_4	B_4	B_4	B_4	B_4	B_4	B_4	B_4	B_3	B_3	B_3										
sp.	Ec |																						PU17									
USNM 301513	Ep |																															
SL 41 mm	R |	S				B	B	B	B	B	B	B	B	B	B	B	B	B	B	B	B	B										
Menidia	En |		-	B_P	B_P	B_3	B_3	B_3	B_3	B_3	B_3	B_3	B_3	B_3	B_3	B_3	B_3	B_3	B_3	B_3	L_3	L_2	L_2	L_2	L_2							
peninsulae	Ec |																									PU16						
BMNH 48.8.6.992	Ep |																															
SL 64 mm	R |	S					B	B	B	B	B	B	B	B	B	B	B	B	B													
Exocoetus	En |		-	B_P	B_P	B_3	B_3	B_3	B_3	B_3	B_3	B_3	B_3	B_3	B_3	B_3	B_3	B_3	B_3	B_3	B_3	B_3	B_3	B_3	B_3	B_3	B_3	B_5	//	B_5		
volitans	Ec |																														PU16	
BMNH 80.1.22.1	Ep |																															
SL 52 mm	R |	S			B	B	B	B	B	B	B	B	B	B	B	B	B	B	B	B	B	B	B	B	LB							
Culaea	En |		-	B_P	B_3	B_4	B_4	B_4	B_4	B_4	B_4	B_4	B_4	B_4	B_4	B_4	B_0	B_0	B_0													
inconstans	Ec |																			PU17												
BMNH 92.12.30.95	Ep |																															
SL 33 mm	R |					B	B	B	B	B	B	B	B	B	B	B	B	B														
Aulichthys	En |		-	B_P	B_3	B_3																										
japonicus	Ec |																										PU27					
BMNH 79.7.20.25	Ep |																															
SL 117 mm	R |						B	B	B	B	B	B	B	B	B	B	B	B	B	B	B	B	B	B	B	B						
Aulorhynchus	En |		-	L_P																												
flavidus	Ec |				PU53																											
BMNH 79.7.10.7	Ep |																															
SL 98 mm	R |																															
Hypoptychus	En |		-	L_P	L_3	L_3																										
dybowskii	Ec |																															PU23
BMNH 79.9.4.14	Ep |																															
SL 64 mm	R |					B	B	B	B	B	B	B	B	B	B	B	B	B	B	B	B	B	B	B	B	B	B	B	//	B	B	B
Morone	En |		-	B_1	B_1	B_4	B_4	B_4	B_4	B_4																						
americana	Ec |															L_5	L_5	L_2	L_2	L_2	L_2	L_2	L_2	L_2	L_2	PU3						
USNM 109851	Ep |																															
SL 44 mm	R |	S				B	B	B	B	B	B	B	B	B	B																	
Pseudanthias	En |		-	B_1	B_1	B_4	B_4	B_4	B_4	B_4	B_4	B_3	B_3	B_5	B_5	B_5	B_5															
squamipinnis	Ec |																	PU12														
BMNH 75.4.5.14	Ep |																															
SL 29 mm	R |	S				B	B	B	B	B	B	B	B																			
Lates	En |		-	B_1	B_1	B_4	B_4	B_4																								
sp.	Ec |												L_3	L_3	L_2	L_2	L_2	L_2	L_2	L_2	L_2	L_2	PU6									
BMNH uncat.	Ep |																															
SL 38 mm	R |	S				B	B	B	B	B	B	B	B	B																		
Lates	En |		-	B_1	B_1	B_4	B_4																									
calcarifer	Ec |														L_5	L_2	L_2	L_2	L_2	L_2	L_2	L_2	PU6									
USNM 44768	Ep |																															
SL 141 mm	R |	S				B	B	B	B	B	B	B	B	B																		
Psammoperca	En |		-	B_1	B_1	B_4	B_4	B_4	B_4	B_4	B_3																					
waigiensis	Ec |												L_3	L_2	L_2	L_2	L_2	L_2	L_2	L_2	L_2	L_2	L_2	L_2	PU4							
USNM 218816	Ep |																															
SL 118 mm	R |	S				B	B	B	B	B	B	B	B	B																		
Doederleinia	En |		-	B_1	B_1	B_4	B_4	B_4	B_4	B_4	B_4	L_4	L_3																			
berycoides	Ec |																L_5	L_5	L_5	L_5	L_5	L_5	L_2	L_2	PU4							
USNM 290474	Ep |																															
SL 127 mm	R |	S				B	B	B	B	B	B	B	B																			

TABLE 8.—Continued.

		Occ	ANA	1	2	3	4	5	6	7	8	9	10	11	12	13	14	15	16	17	18	19	20	21	22	23	24	25	26	//	42
Capros	En |		-		B_4	B_4	B_4	B_4	B_4	B_4	B_3	B_3	B_5	B_5	B_5	$\boldsymbol{B}_5$	L_5	L_5	L_5	L_2	PU5										
aper	Ec |																														
USNM 320065	Ep |																														
SL 57 mm	R |	S			B	$\underline{B}$	$\underline{B}$	$\underline{B}$	$\underline{B}$	$\underline{B}$	$\underline{B}$	$\underline{B}$	B																		
Antigonia	En |		-		B_1	B_4	B_4	B_4	B_4	B_3	B_3	B_3	B_5	B_5	B_5	B_5	L_2	L_2	L_2	L_2	L_2										
sp.	Ec |																					PU4									
USNM 26690	Ep |																														
SL 39 mm	R |	S			B	$\underline{B}$	$\underline{B}$	$\underline{B}$	$\underline{B}$	$\underline{B}$	$\underline{B}$	$\underline{B}$																			
Drepane	En |		-	B_1	B_1	B_4	B_4	B_4	B_3	B_3	B_3	B_3	B_5	B_5																	
punctata	Ec |																L_2	L_2	L_2	L_2	L_2	L_2	PU5								
USNM 267052	Ep |																														
SL 50 mm	R |	S				$\underline{B}$	$\underline{B}$	$\underline{B}$	$\underline{B}$	$\underline{B}$	$\underline{B}$	$\underline{B}$																			
Siganus	En |		-	B_1	B_4	B_4	B_4	B_4	B_4	B_4	B_4	B_3	B_5	L_2	L_2	L_2	L_2	L_2	L_2	L_2											
luridus	Ec |																				PU6										
USNM 218868	Ep |																														
SL 64 mm	R |	S			$\underline{B}$	$\underline{B}$	$\underline{B}$	$\underline{B}$	$\underline{B}$	$\underline{B}$	$\underline{B}$	$\underline{B}$																			
Colisa	En |		-	B_1	B_1	B_4	B_4	B_4	B_4	B_4	B_4	B_4	B_4	B_4	B_3	B_3	B_3	B_2	B_2	B_2	B_2	B_2	$\boldsymbol{B}_2$	L_2	L_2						
soya	Ec |																									PU4					
BMNH 89.2.1.3882	Ep |																														
SL 29 mm	R |	S				$\underline{B}$	$\underline{B}$	$\underline{B}$	$\underline{B}$	$\underline{B}$	$\underline{B}$	$\underline{B}$	$\underline{B}$	$\underline{B}$	$\underline{B}$	$\underline{B}$	$\underline{B}$	$\underline{B}$	$\underline{B}$	$\underline{B}$	B	B	B	B	B						
Trachinus	En |		-	B_1	B_1	B_1	B_1	B_1	B_1	B_2	B_2	B_3	B_4	B_3	B_3	B_5	B_5	B_5	B_5	B_5	B_5	B_5									
vipera	Ec |																						PU16								
BMNH 62.9.30.546	Ep |																														
SL 73 mm	R |	S											$\underline{B}$	B	B	B															
Callionymus	En |		-		B_1	B_1	B_1B_2	B_2	B_2		$\boldsymbol{B}_2$																				
lyra	Ec |											PU12																			
BMNH 91.2.16.1191	Ep |																														
SL ca. 65 mm	R |	S																													
Ectreposebastes	En |		-	B_1	B_1	B_1	B_2	B_4	B_4	$\boldsymbol{B}_4$	B_4																				
imus	Ec |													PU14																	
USNM 198216	Ep |																														
SL 115 mm	R |	S						$\underline{B}$	$\underline{B}$	$\underline{B}$	$\underline{B}$	$\underline{B}$	B																		
Cottus	En |		-	B_1	B_1	B_1	B_3	B_3	B_3	B_3	B_3	B_3	B_3	B_3	B_5	B_5	B_5	B_5	B_5	B_5											
sp.	Ec |																				PU16										
USNM 280751	Ep |																														
SL 53 mm	R |	S										$\underline{B}$	$\underline{B}$	B																	
Hexagrammus	En |		-	B_1	B_1	B_4	B_4	B_4	B_4	B_4	B_4	B_4	B_4	B_4	B_4	B_4	B_3	B_3	B_3	B_3	B_3	B_3	$\boldsymbol{B}_3$	$\boldsymbol{B}_3$							
decagrammus	Ec |																									PU33					
USNM 104656	Ep |																														
SL 46 mm	R |			S		B	$\underline{B}$	$\underline{B}$	$\underline{B}$	$\underline{B}$	$\underline{B}$	$\underline{B}$	$\underline{B}$	$\underline{B}$	$\underline{B}$	$\underline{B}$	$\underline{B}$	$\underline{B}$	$\underline{B}$	$\underline{B}$	$\underline{B}$	$\underline{B}$	$\underline{B}$	B	B						
Sphyraena	En |		-	B_1	B_1	B_4																									
barracuda	Ec |											L_3	L_3	L_3	L_2	L_2	L_2	L_2	PU9												
USNM uncat.	Ep |																														
SL 14 mm	R |	S				$\underline{B}$	$\underline{B}$	$\underline{B}$	$\underline{B}$	$\underline{B}$	$\underline{B}$	$\underline{B}$	$\underline{B}$	B																	
Sphyraena	En |		-	B_1	B_1	B_4	B_4	B_4	B_4	$\boldsymbol{B}_4$																					
barracuda	Ec |										L_3	L_3	L_3	L_3	L_2	L_2	L_2	L_2	L_2	L_2	L_2	L_2	PU5								
USNM 2226399	Ep |																														
SL 79 mm	R |	S				$\underline{B}$	$\underline{B}$	$\underline{B}$	$\underline{B}$	$\underline{B}$	$\underline{B}$	$\underline{B}$	$\underline{B}$	B																	
Caranx	En |		-	B_1	B_1	B_4	B_4	B_4	B_4	B_4	B_3	B_3	B_3	B_5	B_2	B_2	B_2	B_2	B_2	L_2	L_2	L_2	L_2	L_2	L_2	PU3					
chrysos	Ec |																														
USNM 104656	Ep |																														
SL 46 mm	R |	S				$\underline{B}$	$\underline{B}$	$\underline{B}$	$\underline{B}$	$\underline{B}$	$\underline{B}$	$\underline{B}$																			
Rachycentron	En |		-	B_1	B_1	B_4	B_4	B_4	B_4	B_4	B_4	B_3	B_3	B_3	B_5	B_2	L_2														
canadum	Ec |																	L_2	L_2	L_2	L_2	L_2	L_2	L_2	L_2	PU3					
USNM 120079	Ep |																														
SL 148 mm	R |	S			$\underline{B}$	$\underline{B}$	$\underline{B}$	$\underline{B}$	$\underline{B}$	$\underline{B}$	$\underline{B}$	$\underline{B}$	$\underline{B}$	B																	
Scomberomorus	En |	B	-	B_1	B_1	B_3	B_4	B_4	B_3	B_3	B_3	B_3	B_3	B_3	B_3	B_3	B_3	B_2	B_2	B_2	B_2	B_2	B_2	B_2							
tritor	Ec |													L_3	L_3	L_3	L_3	L_2	L_2	L_2	L_2	L_2	L_2	L_2	L_2	L_2	L_2	L_2	L_2	//	PU5
USNM 193839	Ep |																														
SL 144 mm	R |	S				B	B	B	B	B	B	B	B	B	B	B	B	B	B	B	B	B									
Lepadogaster	En |		-		B_2	B_2	B_2	B_2	B_2	B_2	B_2	B_2	B_2	B_2	B_2	B_2	B_2	B_3	B_5	B_5	B_5										
lepadogaster	Ec |																					PU13									
USNM 276289	Ep |																														
SL 36 mm	R |				B_N	B_N	B_N	B_N	B_N	B_N	B_N	B_N	B_N	B_N	B_0	B_0	B_0														

TABLE 8.—Continued.

		Occ	ANA	1	2	3	4	5	6	7	8	9	10	11	12	13	14	15	16	17	18	19	20	21	22	23	24	25	26	27	28	29
Micropercops	En |		-	$\mathbf{B}_1$	$\mathbf{B}_1$	$\mathbf{B}_4$	$\mathbf{B}_4$	$\mathbf{B}_4$	$\mathbf{B}_4$	$\mathbf{B}_4$	$\mathbf{B}_4$	$\mathbf{B}_4$	$\mathbf{B}_4$		$\mathbf{B}_0$																	
sp.	Ec |																			**PU17**												
AMS I.27275-002	Ep |																															
SL 42 mm	R |	S				B	B	B	B	B	B	B	B	B	B	B	B	B	$\underline{B}$													
Upeneus	Nn |				L_1	L_1	L_1	L_1	L_1	L_1	L_1	L_1	L_1	L_1	L_1	L_1	L_1															
tragula	En |		-	$\mathbf{B}_1$	$\mathbf{B}_3$	$\mathbf{B}_4$	$\mathbf{B}_4$	$\mathbf{B}_4$	$\mathbf{B}_4$	$\mathbf{B}_4$	$\mathbf{B}_5$	$\mathbf{B}_5$	$\mathbf{B}_5$	$\mathbf{B}_5$	$\mathbf{B}_5$	$\mathbf{B}_2$	L_2	L_2	L_2	L_2	L_2	L_2	L_2	L_2	L_2	L_2	**PU1**					
BMNH uncat.	Ep |																															
SL 78 mm	R |	S				$\underline{B}$	$\underline{B}$	$\underline{B}$	$\underline{B}$	$\underline{B}$	$\underline{B}$	$\underline{B}$	B																			
Dascyllus	Nn |				L_1	L_1	L_1	L_1	L_1	L_1	L_1	L_1	L_1	L_1	L_1	L_1	L_1	L_1	L_1	L_1	L_1											
aruanus	En |		-	$\mathbf{B}_1$	$\mathbf{B}_1$	$\mathbf{F}_4$	$\mathbf{F}_4$	$\mathbf{F}_4$	$\mathbf{F}_4$	$\mathbf{F}_4$	$\mathbf{F}_4$	$\mathbf{F}_4$	$\mathbf{F}_4$	$\mathbf{F}_4$	$\mathbf{B}_5$	$\mathbf{B}_5$	$\mathbf{B}_5$	$\mathbf{B}_5$	$\mathbf{B}_5$			**PU8**										
BMNH uncat.	Ep |																															
SL 25 mm	R |	S				$\underline{B}$	$\underline{B}$	$\underline{B}$	$\underline{B}$	$\underline{B}$	$\underline{B}$	$\underline{B}$	$\underline{B}$	$\underline{B}$																		
Ammodytes	Nn |							L_3	L_2	L_2	L_2	L_2	L_2	L_2	L_2	L_2	L_2	L_2	L_2	L_2	L_2	L_2	L_2	L_2	L_2	L_2	L_2	L_2	L_2	L_2	L_2	L_2
tobianus	En |		-	$\mathbf{B}_1$	$\mathbf{B}_1$	$\mathbf{B}_4$	$\mathbf{B}_4$	$\mathbf{B}_4$	$\mathbf{B}_4$	$\mathbf{B}_4$	$\mathbf{B}_4$	$\mathbf{B}_4$	$\mathbf{B}_4$	$\mathbf{B}_4$	$\mathbf{B}_4$	$\mathbf{B}_4$	$\mathbf{B}_4$	$\mathbf{B}_4$	$\mathbf{B}_4$	$\mathbf{B}_4$	$\mathbf{B}_4$	$\mathbf{B}_4$	$\mathbf{B}_4$	$\mathbf{B}_4$	$\mathbf{B}_4$	$\mathbf{B}_4$	$\mathbf{B}_4$	$\mathbf{B}_4$	$\mathbf{B}_4$	$\mathbf{B}_4$	$\mathbf{B}_4$	$\mathbf{B}_4$
BMNH uncat.	Ep |																															
SL 93 mm	R |	S				B	B	B	B	B	B	B	B	B	B	B	B	B	B	B	B	B	B	B	B	B	B	B	B	B	B	B
Polydactylus	Nn |			L_1	L_1	L_1	L_2	L_2	L_2	L_2	L_2	L_1	L_1	L_1	L_1																	
virginicus	En |		-	$\mathbf{B}_1$	$\mathbf{B}_2$	$\mathbf{B}_4$	$\mathbf{B}_4$	$\mathbf{B}_4$	$\mathbf{B}_3$	$\mathbf{B}_3$	L_4	L_4	L_4	L_5	L_5	L_5	L_5	L_5														
BMNH 67.8.16.312	Ec |												L_2	L_2	L_2	L_2	L_2	L_2	L_2	L_2	L_2	L_2	L_2	L_2	**PU3**							
SL ca. 100 mm	R |	S				$\underline{B}$	$\underline{B}$	$\underline{B}$	$\underline{B}$	$\underline{B}$	$\underline{B}$	$\underline{B}$	$\underline{B}$																			
Samaris	Nn |			$L_1{}_{<}B_1$	$_{>}B1$	$_{>}B_1$	$_{x}V_0$	$_{x}V_0$	$_{x}V_0$	$_{x}V_0$	$_{x}V_0$	$_{<}V_0$	$_{<}V_0$	$_{<}V_0$	$_{<}V_0$	$_{<}V_0$	$_{>}V_0$	V_0	V_0	V_0	V_0	V_0	V_0	V_0	V_0	V_0	$_{>}V_0$	$_{<}V_0$	V_0	V_0	$_{<}V_0$	$_{x}V_0$
cristatus	En |		-	$\mathbf{B}_1$	$\mathbf{B}_1$	$\mathbf{B}_2$	$\mathbf{B}_3$	$\mathbf{B}_3$	$\mathbf{B}_3$	$\mathbf{B}_3$	$\mathbf{B}_3$	$\mathbf{B}_3$																				
UPRM uncat.	Ep |																															
SL 102 mm	R |	S			B	B	B	$_{<}B$	$_{<}B$	B	B	$_{<D}B$	$_{>D}B$	$_{>}D_0$	$_{<}D_0$	D_0	D_0	$_{<}D_0$	D_0	$_{<}D_0$	D_0	D_0	$_{<}D_0$	D_0	$_{>}D_0$	$_{>}D_0$	D_0	$_{>}D_0$	D_0	D_0	D_0	D_0
Bothus	Nn |	*B*		B_1	$_{<}B_1$		$_{x}B_1$	$_{x}B_1$	$_{x}B_1$	$_{<v}B_1$	$_{<v}B_1$	$_{<v}B_1$	$_{<v}B_1$	$_{x}V_0$	$_{x}V_0$	$_{x}V_0$	$_{x}V_0$	$_{x}V_0$	$_{x}V_0$	$_{x}V_0$	$_{x}V_0$	$_{x}V_0$	$_{x}V_0$	$_{x}V_0$	$_{x}V_0$	$_{x}V_0$	$_{x}V_0$	$_{x}V_0$	$_{x}V_0$	$_{x}V_0$	$_{x}V_0$	$_{x}V_0$
podus	En |		-		$\mathbf{B}_2$	$\mathbf{B}_2$	$\mathbf{B}_2$	$\mathbf{B}_2$	$\mathbf{B}_2$	$\mathbf{B}_2$	$\mathbf{B}_2$	$\mathbf{B}_2$																				
USNM 261536	Ep |																															
SL 92 mm	R |	S			B	B	B	B	$_{<}B$	$_{<D}B$	$_{<D}B$	$_{<D}B$	$_{<D}B$	$_{x}D_0$	$_{x}D_0$	$_{x}D_0$	$_{x}D_0$	$_{x}D_0$	$_{x}D_0$	$_{x}D_0$	$_{x}D_0$	$_{x}D_0$	$_{x}D_0$	$_{x}D_0$	$_{x}D_0$	$_{x}D_0$	$_{x}D_0$	$_{x}D_0$	$_{x}D_0$	$_{x}D_0$	$_{x}D_0$	$_{x}D_0$

TABLE 8.—Continued.

	30	31	32	33	34	35	36	37	38	39	40	41	42	43
Ammodyt	L_2	L_2	L_2	L_2	L_1	L_1	L_1	L_1	L_1	L_1	L_1			
	B_4	**B_4**	**B_3**	**B_3**	**B_3**	**B_2**	**B_2**	**B_2**	**B_2**	**B_2**	L_2	L_2	L_2	**PU20**
	B	B	B	B	B									
Samaris	x^{V}_{0}	x^{V}_{0}												
			PU8											
	$>^{D}_{0}$													
Bothus	x^{V}_{0}	x^{V}_{0}	x^{V}_{0}	x^{V}_{0}	x^{V}_{0}	x^{V}_{0}	x^{V}_{0}	x^{V}_{0}						
									PU2					
	x^{D}_{0}	x^{D}_{0}	x^{D}_{0}	x^{D}_{0}	x^{D}_{0}	x^{D}_{0}	x^{D}_{0}	x^{D}_{0}						

Appendix 1

Alphabetical List of Neopterygian Genera Cited in Text and Tables

(Names in bold italic indicate genera for which we have studied the intermusculars; those in italic indicate genera studied by others. References to Tables 1–5, 7, and 8 are given so that the list provides an alphabetical index to those tables. Family names are given except where a genus is the type of a family. Names of extinct genera are preceded by a dagger (†), and names of extinct paraphyletic stem-group families are placed in quotation marks.)

Abudefduf, Pomacentridae	
Acanthopsoides, Cobitidae	Table 4
Adioryx, Holocentridae	
Agonostomus, Mugilidae	Table 8
Ahliesaurus, Notosudidae	Table 5
Albula	Table 3
Alepisaurus	Table 5
Alepocephalus	
Alestes, Characidae	Table 4
†***Allothrissops***	
Alosa, Clupeidae	
Amia	
Ammodytes	Table 8
†***Anaethalion***, Elopomorpha incertae sedis	
Anchoa, Engraulidae	Table 3
Anguilla	Table 3
Anodontostoma, Clupeidae	
Anomalops	Table 7
Anoplogaster	Table 7
Anoplopoma	
Anotopterus	
Antigonia, Caproidae	Table 8
Aphredoderus	Table 7
Aplochiton, Galaxiidae	
Arapaima	
Argentina	Table 4
Ariosoma, Congridae	
†***Ascalabos***, "Leptolepididae"	
†*Asiatolepis*, Osteoglossoidei incertae sedis	
Astroscopus, Uranoscopidae	
Ateleopus	
Aulichthys, Aulorhynchidae	Table 8
Aulopus	Table 5
Aulorhynchus	Table 8
Aulostomus	
Barbourisia	Table 7
Bathypterois, Ipnopidae	Table 5
Bathysaurus	Table 5
Bathytroctes, Alepocephalidae	
Bedotia, Atherinidae	Table 8
Bembrops, Percophididae	
Binghamichthys, Alepocephalidae	
Bothus	Table 8
Brevoortia, Clupeidae	
Callionymus	Table 8
Capros	Table 8
Caranx	Table 8
Centroberyx, Berycidae	Table 7
Centropomus	
Cetengraulis, Engraulidae	

Champsodon	
Chanos	Table 4
Chiasmodon	
Chirocentrodon, Pellonidae	
Chirocentrus	Table 3
Chlorophthalmus	Table 5
Clupanodon, Clupeidae	
Clupea	Table 3
Clupeichthys, Clupeidae	Table 3
Coccorella, Evermannellidae	Table 5
Coilia	
Colisa, Belontiidae	Table 8
Cookeolus, Priacanthidae	
Coregonus	
Cottus	Table 8
Cromeria, Kneriidae	
Culaea, Gasterosteidae	Table 8
Cynothrissa, Clupeidae	
Cyprinus	
Cyttopsis, Zeidae	
Dactyloscopus	
Dallia, Umbridae	
Dascyllus, Pomacentridae	Table 8
Denticeps	Table 3
Diplophos, Gonostomatidae	Table 5
Diretmus	Table 7
Doederleinia, Acropomatidae	Table 8
Dorosoma, Clupeidae	
Drepane	Table 8
Dysalotus, Chiasmodontidae	
Echeneis	
Ectreposebastes, Scorpaenidae	Table 8
Eigenmannia, Sternopygidae	
Elassoma	Table 8
Elops	Table 3
Engraulis	
†***Eoaulostomus***, Aulostomidae	
Epigonus	
Esox	Table 4
Exocoetus	Table 8
Etrumeus, Clupeidae	
Euthynnus, Scombridae	
Evermanella	
Forsterygion, Tripterygiidae	
Gadus	
Gaidropsarus, Phycidae	Table 7
Galaxias	Table 4
Gibberichthys	Table 7
Gigantura	Table 5
Glossanodon, Argentinidae	Table 4

Gnathagnus, Uranoscopidae	
Gnatholepis, Gobiidae	
Gonorynchus	Tables 2,4
Grasseichthys, Kneriidae	
Gymnarchus	
Gymnotus	
Halosaurus	Table 3
Harengula, Clupeidae	
Harpadon	Table 5
Hemerocoetes, Percophididae	
Hemibarbus, Cyprinidae	Table 4
Heterotis, Arapaimidae	Table 3
Hexagrammus	Table 8
Hiodon	Table 3
Hispidoberyx	Table 7
Holocentrus	Table 7
Hoplichthys	
Hoplostethus, Trachichthyidae	Table 7
†*Huashia*	
†*Hulettia,* Halecostomi incertae sedis	
Hypentelium, Catostomidae	Table 4
Hypomesus, Osmeridae	Table 4
Hypoptychus, Aulorhynchidae	Table 8
Hypopygus, Hypopomidae	
Ilisha, Pellonidae	
Indostomus	
Isichthys, Mormyridae	
Jenkinsia, Clupeidae	Table 3
†*Jiaohichthys,* Osteoglossomorpha incertae sedis	
†***Jungersenichthys,*** Aulostomidae	
Kathetostoma, Uranoscopidae	
Katsuwonus, Scombridae	
Kneria	
†*Kuntulunia,* Huashiidae	
Lampris	Table 7
Lates, Centropomidae	Table 8
Lepadogaster, Gobiesocidae	Table 8
Lepidogalaxias	Table 4
Lepisosteus	
Leptochilichthys, Alepocephalidae	
Leptoderma, Alepocephalidae	
†***Leptolepis***	
Leptoscopus	
Lile, Clupeidae	
Liza, Mugilidae	Table 8
Lovettia, Galaxiidae	
†*Lycoptera,* Hiodontidae	
Macroparalepis, Paralepididae	Table 5
Mallotus, Osmeridae	
Marcusenius, Mormyridae	
Maurolicus, Sternoptychidae	Table 5
Megalops	Table 3
Melamphaes	
Menidia, Atherinidae	Table 8
Metavelifer, Veliferidae	Table 7
Micropercops, Odontobutidae	Table 8
Monocentris	Table 7
Morone	Table 8
Mulloidichthys, Mullidae	
Myctophum	Table 5
Myripristis, Holocentridae	
Nematalosa, Clupeidae	
Neoscopelus	Table 5
Notacanthus	Table 3
Notopterus	
Novumbra, Umbridae	Table 4
Odontognathus, Pristigasteridae	
Omosudis	
Oncorhynchus, Salmonidae	
Opisthonema, Clupeidae	
Opisthoproctus, Bathylagidae	
Opisthopterus, Pristigasteridae	
Osmerus	Table 4
Osteoglossum	Table 3
Ostichthys, Holocentridae	
Pantodon	
Papyrocranus, Notopteridae	
Paragalaxias, Galaxiidae	
Parakneria, Kneriidae	Table 4
Paralepis	Table 5
†*Paralycoptera,* Osteoglossoidei incertae sedis	
Parapercis, Pinguipedidae	
Parasudis, Chlorophthalmidae	Table 5
Parazen	Table 7
Pegasus	
Pellona	
Pellonula, Clupeidae	
Perca	
Percopsis	Table 7
†***Pholidolepis,*** "Pholidophoridae"	
†***Pholidophorus***	
Photoblepharon, Anomalopidae	Table 7
Phractolaemus, Kneriidae	
Platycephalus	
Searsia, Alepocephalidae	Table 4
Plecoglossus, Osmeridae	Table 4
†*Plesiolycoptera,* Hiodontidae	
Pollichthys, Photichthyidae	Table 5
Polydactylus, Polynemidae	Table 8
Polymixia	Tables 1, 7
Poromitra, Melamphaidae	Table 7
†***Proleptolepis,*** "Leptolepididae"	
Prosopium, Coregonidae	
Prototroctes, Retropinnidae	
Psammoperca, Centropomidae	Table 8
Psettodes	
Pseudanthias, Serranidae	Table 8
Pseudotrichonotus	Table 5
Pterophryne, Antennariidae	
Rachycentron, Coryphaenidae	Table 8
Raniceps	Table 7
Retropinna	
Rinoctes, Alepocephalidae	
Rondeletia	Table 7
Salangichthys, Salangidae	Table 4
Salmo	Table 4
Salvelinus, Salmonidae	
Samaris	Table 8
Sarda, Scombridae	
Sardinella, Clupeidae	
Sardinops, Clupeidae	
Sargocentron, Holocentridae	
Saurida, Harpadontidae	Table 5
Scomber	
Scomberomorus, Scombridae	Table 8
Scopelarchoides, Scopelarchidae	Table 5
Scopelengys, Neoscopelidae	
Scopelogadus, Melamphaidae	

Appendix 2

Outline Classification of Neopterygians Including Genera Cited in Text and Tables

(Names in bold italic indicate genera for which we have studied the intermusculars; those in italic indicate genera studied by others. References to Tables 1–5, 7, and 8 are given so that the classification provides a systematic index to those tables. Classification follows Nelson (1984), Johnson and Gill (1994), Johnson and Patterson (1993), Patterson (1993), and some results of this paper. Extinct taxa are preceded by a dagger (†), and names of extinct paraphyletic stem-group families are placed in quotation marks.)

- NEOPTERYGII
  - LEPISOSTEIFORMES
    - LEPISOSTEIDAE
      - ***Lepisosteus***
- HALECOSTOMI
  - *incertae sedis*
    - †Huletta
  - AMIIFORMES
    - AMIIDAE
      - ***Amia***
- TELEOSTEI
  - †"PHOLIDOPHORIDAE"
    - ***Pholidophorus***
    - ***Pholidolepis***
  - †"LEPTOLEPIDIDAE"
    - ***Proleptolepis***
    - ***Leptolepis***
    - ***Ascalabos***
    - ***Tharsis***
    - *Todiltia*
  - †ICHTHYODECTIFORMES
    - ***Allothrissops***
    - ***Thrissops***
  - OSTEOGLOSSOMORPHA
    - *incertae sedis*
      - †*Jiaohichthys*
    - †HUASHIIDAE
      - *Huashia*
      - *Kuntulunia*
    - OSTEOGLOSSIFORMES
      - NOTOPTEROIDEI
        - HIODONTIDAE
          - ***Hiodon*** Table 3
          - †*Lycoptera*
          - †*Plesiolycoptera*
          - †*Yanbiania*
        - NOTOPTERIDAE
          - ***Notopterus***
          - ***Papyrocranus***
          - ***Xenomystus*** Table 3
        - MORMYRIDAE
          - ***Isichthys***
          - ***Marcusenius***
        - GYMNARCHIDAE
          - ***Gymnarchus***
      - OSTEOGLOSSOIDEI
        - *incertae sedis*
          - †*Asiatolepis*
          - †*Paralycoptera*
        - PANTODONTIDAE
          - *Pantodon*
        - OSTEOGLOSSIDAE
          - ***Osteoglossum*** Table 3
        - ARAPAIMIDAE
          - ***Arapaima***
          - ***Heterotis*** Table 3
  - ELOPOMORPHA
    - *incertae sedis*
      - †***Anaethalion***
    - ELOPIFORMES
      - ELOPIDAE
        - ***Elops*** Table 3
    - MEGALOPIFORMES
      - MEGALOPIDAE
        - ***Megalops*** Table 3
    - ANGUILLIFORMES
      - ALBULIDAE
        - ***Albula*** Table 3
      - HALOSAURIDAE
        - ***Halosaurus*** Table 3
      - NOTACANTHIDAE
        - ***Notacanthus*** Table 3
      - ANGUILLIDAE
        - ***Anguilla*** Table 3
      - CONGRIDAE
        - *Ariosoma*
  - CLUPEOMORPHA
    - CLUPEIFORMES
      - DENTICIPITOIDEI
        - DENTICIPITIDAE
          - ***Denticeps*** Table 3
      - CLUPEOIDEI
        - ENGRAULOIDEA
          - COILIIDAE
            - ***Coilia***
            - ***Setipinna***
            - ***Thryssa***
          - ENGRAULIDAE
            - ***Anchoa*** Table 3
            - ***Cetengraulis***
            - ***Engraulis***
            - ***Stolephorus***
        - PRISTIGASTEROIDEA
          - PELLONIDAE
            - ***Chirocentrodon***
            - ***Ilisha***
            - ***Pellona***
          - PRISTIGASTERIDAE
            - ***Odontognathus***
            - ***Opisthopterus***
        - CLUPEOIDEA
          - CHIROCENTRIDAE
            - ***Chirocentrus*** Table 3
          - CLUPEIDAE
            - DUSSUMIERIINAE
              - ***Etrumeus***
              - ***Jenkinsia*** Table 3
              - ***Spratelloides***
            - PELLONULINAE
              - ***Clupeichthys*** Table 3
              - ***Cynothrissa***
              - ***Pellonula***
              - ***Sierrathrissa***
            - ALOSINAE
              - ***Alosa***
              - ***Brevoortia***
            - CLUPEINAE
              - ***Clupea*** Table 3
              - ***Harengula***
              - *Lile*
              - ***Opisthonema***
              - *Sardinella*
              - *Sardinops*
              - ***Sprattus***
            - DOROSOMATINAE
              - ***Anodontostoma***
              - *Clupanodon*
              - ***Dorosoma***
              - *Nematalosa*
  - OSTARIOPHYSI
    - ANOTOPHYSI
      - GONORYNCHIFORMES
        - CHANIDAE
          - ***Chanos*** Table 4
        - GONORYNCHIDAE
          - ***Gonorynchus*** Tables 2, 4
        - KNERIIDAE
          - ***Cromeria***
          - ***Grasseichthys***
          - *Kneria*
          - ***Parakneria*** Table 4
          - ***Phractolaemus***

OTOPHYSI
CYPRINIFORMES
CYPRINIDAE
Cyprinus
Hemibarbus Table 4
COBITIDAE
Acanthopsoides Table 4
CATOSTOMIDAE
Hypentelium Table 4
CHARACIFORMES
CHARACIDAE
Alestes Table 4
SILURIFORMES
GYMNOTOIDEI
STERNOPYGIDAE
Eigenmannia
Sternopygus
HYPOPOMIDAE
Hypopygus
GYMNOTIDAE
Gymnotus
EUTELEOSTEI
ESOCIFORMES
ESOCIDAE
Esox Table 4
UMBRIDAE
Dallia
Novumbra Table 4
Umbra Table 4
SALMONIFORMES
ARGENTINOIDEI
ARGENTINOIDEA
ARGENTINIDAE
Argentina Table 4
Glossanodon Table 4
BATHYLAGIDAE
Opisthoproctus
ALEPOCEPHALOIDEA
ALEPOCEPHALIDAE
Alepocephalus
Bathytroctes
Binghamichthys
Leptochilichthys
Leptoderma
Rinoctes
Searsia Table 4
Talismania
SALMONOIDEI
COREGONIDAE
Coregonus
Prosopium
Stenodus
SALMONIDAE
Oncorhynchus
Salmo Table 4
Salvelinus
Thymallus Table 4
OSMEROIDEI
OSMEROIDEA
OSMERIDAE
Hypomesus Table 4
Mallotus
Osmerus Table 4
Plecoglossus Table 4
Spirinchus
SALANGIDAE
Salangichthys Table 4
GALAXIOIDEA
RETROPINNIDAE
Prototroctes
Retropinna
Stokellia Table 4
LEPIDOGALAXIIDAE
Lepidogalaxias Table 4
GALAXIIDAE
Aplochiton
Galaxias Table 4
Lovettia
Paragalaxias
NEOTELEOSTEI
incertae sedis
ATELEOPODIDAE
Ateleopus
STOMIIFORMES
GONOSTOMATIDAE
Diplophos Table 5
Triplophos
PHOTICHTHYIDAE
Pollichthys Table 5
STERNOPTYCHIDAE
Maurolicus Table 5
AULOPIFORMES
incertae sedis
IPNOPIDAE
Bathypterois Table 5
CHLOROPHTHALMIDAE
Chlorophthalmus Table 5
Parasudis Table 5
SYNODONTOIDEI
AULOPIDAE
Aulopus Table 5
PSEUDOTRICHONOTIDAE
Pseudotrichonotus Table 5
SYNODONTIDAE
Synodus Table 5
Trachinocephalus Table 5
HARPADONTIDAE
Harpadon Table 5
Saurida Table 5
ALEPISAUROIDEI
EVERMANELLIDAE
Evermanella
Coccorella Table 5
NOTOSUDIDAE
Ahliesaurus Table 5
Scopelosaurus Table 5
SCOPELARCHIDAE
Scopelarchoides Table 5
PARALEPIDIDAE
Macroparalepis Table 5
Paralepis Table 5
ALEPISAURIDAE
Alepisaurus Table 5
ANOTOPTERIDAE
Anotopterus
BATHYSAURIDAE
Bathysaurus Table 5
GIGANTURIDAE
Gigantura Table 5
OMOSUDIDAE
Omosudis
MYCTOPHIFORMES
MYCTOPHIDAE
Myctophum Table 5
NEOSCOPELIDAE
Neoscopelus Table 5
Scopelengys
Solivomer
ACANTHOMORPHA
LAMPRIDIFORMES
VELIFERIDAE
Metavelifer Table 7
Velifer Table 7
LAMPRIDIDAE
Lampris Table 7
POLYMIXIIFORMES
POLYMIXIIDAE
Polymixia Tables 1, 7
PARACANTHOPTERYGII
PERCOPSIFORMES
PERCOPSIDAE
Percopsis Table 7
APHREDODERIDAE
Aphredoderus Table 7
GADIFORMES
RANICIPITIDAE
Raniceps Table 7
PHYCIDAE
Gaidropsarus Table 7
GADIDAE
Gadus
LOPHIIFORMES
ANTENNARIIDAE
Pterophryne
STEPHANOBERYCIFORMES
MELAMPHAIDAE
Melamphaes
Poromitra Table 7
Scopelogadus
STEPHANOBERYCIDAE
Stephanoberyx Table 7
GIBBERICHTHYIDAE
Gibberichthys Table 7
HISPIDOBERYCIDAE
Hispidoberyx Table 7
BARBOURISIIDAE
Barbourisia Table 7
RONDELETIIDAE
Rondeletia Table 7
BERYCIFORMES
HOLOCENTRIDAE
Adioryx
Holocentrus Table 7
Myripristis
Ostichthys
Sargocentron
BERYCIDAE
Centroberyx Table 7
TRACHICHTHYIDAE
Hoplostethus Table 7
Trachichthys Table 7
MONOCENTRIDAE
Monocentris Table 7

ANOMALOPIDAE
- ***Anomalops*** Table 7
- ***Photoblepharon*** Table 7

DIRETMIDAE
- ***Diretmus*** Table 7

ANOPLOGASTRIDAE
- ***Anoplogaster*** Table 7

ZEIFORMES

PARAZENIDAE
- ***Parazen*** Table 7

ZENIONTIDAE
- ***Zenion***

GRAMMICOLEPIDIDAE
- ***Xenolepidichthys*** Table 7

ZEIDAE
- ***Cyttopsis***

PERCOMORPHA

SMEGMAMORPHA

ELASSOMATIDAE
- ***Elassoma*** Table 8

MUGILIDAE
- ***Agonostomus*** Table 8
- ***Liza*** Table 8

AULORHYNCHIDAE
- ***Aulichthys*** Table 8
- ***Aulorhynchus*** Table 8
- ***Hypoptychus*** Table 8

INDOSTOMIDAE
- ***Indostomus***

PEGASIDAE
- ***Pegasus***

GASTEROSTEIDAE
- ***Culaea*** Table 8

AULOSTOMIDAE
- ***Aulostomus***
- †***Eoaulostomus***
- †***Jungersenichthys***
- †***Synhypuralis***

ATHERINOMORPHA

ATHERINIDAE
- ***Bedotia*** Table 8
- ***Menidia*** Table 8

EXOCOETIDAE
- ***Exocoetus*** Table 8

SCORPAENIFORMES

SCORPAENIDAE
- ***Ectreposebastes*** Table 8

PLATYCEPHALIDAE
- ***Platycephalus***

HOPLICHTHYIDAE
- ***Hoplichthys***

ANOPLOPOMATIDAE
- ***Anoplopoma***

HEXAGRAMMIDAE
- ***Hexagrammus*** Table 8

COTTIDAE
- ***Cottus*** Table 8

PERCIFORMES

PERCOIDEI

CENTROPOMIDAE
- ***Centropomus***
- ***Lates*** Table 8
- ***Psammoperca*** Table 8

MORONIDAE
- ***Morone*** Table 8

ACROPOMATIDAE
- ***Doederleinia*** Table 8

SERRANIDAE
- ***Pseudanthias*** Table 8
- ***Serranus***

PERCIDAE
- *Perca*

EPIGONIDAE
- ***Epigonus***

CAPROIDAE
- ***Antigonia*** Table 8
- ***Capros*** Table 8

PRIACANTHIDAE
- ***Cookeolus***

MULLIDAE
- ***Mulloidichthys***
- ***Upeneus*** Table 8

CARANGOIDEI

CARANGIDAE
- ***Caranx*** Table 8

ECHENEIDIDAE
- ***Echeneis***

CORYPHAENIDAE
- ***Rachycentron*** Table 8

POLYNEMOIDEI

POLYNEMIDAE
- ***Polydactylus*** Table 8

LABROIDEI

POMACENTRIDAE
- ***Abudefduf***
- ***Dascyllus*** Table 8

TRACHINOIDEI

TRICHODONTIDAE
- ***Trichodon***

PINGUIPEDIDAE
- ***Parapercis***

PERCOPHIDIDAE
- ***Bembrops***
- ***Hemerocoetes***

CHIASMODONTIDAE
- ***Chiasmodon***
- ***Dysalotus***

CHAMPSODONTIDAE
- ***Champsodon***

LEPTOSCOPIDAE
- ***Leptoscopus***

AMMODYTIDAE
- ***Ammodytes*** Table 8

TRACHINIDAE
- ***Trachinus*** Table 8

URANOSCOPIDAE
- ***Astroscopus***
- ***Gnathagnus***
- ***Kathetostoma***
- *Selenoscopus*
- ***Uranoscopus***

BLENNIOIDEI

TRIPTERYGIIDAE
- ***Forsterygion***

DACTYLOSCOPIDAE
- ***Dactyloscopus***

CALLIONYMOIDEI

CALLIONYMIDAE
- ***Callionymus*** Table 8
- ***Synchiropus***

GOBIESOCOIDEI

GOBIESOCIDAE
- ***Lepadogaster*** Table 8

GOBIOIDEI

ODONTOBUTIDAE
- ***Micropercops*** Table 8

XENISTHMIDAE
- *Rotuma*

GOBIIDAE
- ***Gnatholepis***

ACANTHUROIDEI

DREPANIDAE
- ***Drepane*** Table 8

SIGANIDAE
- ***Siganus*** Table 8

SCOMBROIDEI

SPHYRAENIDAE
- ***Sphyraena*** Table 8

SCOMBRIDAE
- *Euthynnus*
- *Katsuwonus*
- *Sarda*
- ***Scomber***
- ***Scomberomorus*** Table 8
- *Thunnus*

ANABANTOIDEI

BELONTIIDAE
- ***Colisa*** Table 8

PLEURONECTIFORMES

PSETTODIDAE
- ***Psettodes***

BOTHIDAE
- ***Bothus*** Table 8

SAMARIDAE
- ***Samaris*** Table 8

Literature Cited

Allis, E.P.

1898. The Homologies of the Occipital and First Spinal Nerves of *Amia* and Teleosts. *Zoological Bulletin,* 2:83–97.

1903. The Skull, and the Cranial and First Spinal Muscles and Nerves in *Scomber scomber. Journal of Morphology,* 18:45–328.

1909. The Cranial Anatomy of the Mail-cheeked Fishes. *Zoologica,* Stuttgart, 22(57): iii + 1–219.

Amaoka, K.

1969. Studies on the Sinstral Flounders Found in the Waters around Japan—Taxonomy, Anatomy and Phylogeny. *Journal of the Shimonoseki University of Fisheries,* 18:65-340.

Arratia, G.

1987. *Anaethalion* and Similar Teleosts (Actinopterygii, Pisces) from the Late Jurassic (Tithonian) of Southern Germany and Their Relationships. *Palaeontographica,* A200:1–44.

Arratia, G., and H.-P. Schultze

1992. Reevaluation of the Caudal Skeleton of Certain Actinopterygian Fishes, III: Salmonidae, Homologization of Caudal Skeletal Structures. *Journal of Morphology,* 214:187–249.

Baldwin, C.C., and G.D. Johnson

In press. A Larva of the Atlantic Flashlight Fish, *Kryptophanaron alfredi* (Beryciformes: Anomalopidae), with a Comparative Look at Known Beryciform and Stephanoberyciform Larvae. *Bulletin of Marine Science,* volume 55.

Baudelot, É.

1868. Considérations relatives à la pièce scapulaire des silures. *Bulletin de la Société des Sciences Naturelles de Strasbourg,* 12:83–84.

Begle, D.P.

1991. Relationships of the Osmeroid Fishes and the Use of Reductive Characters in Phylogenetic Analysis. *Systematic Zoology,* 40:33–53.

1992. Monophyly and Relationships of the Argentinoid Fishes. *Copeia,* 1992:350–366.

Bruch, C.W.L.

1861. *Vergleichende Osteologie des Rheinlachses (Salmo salar L.) mit besonderer Berücksichtigung der Myologie nebst einleitenden Bemerkungen über die skelettbildenden Gewebe der Wirbelthiere.* 22 pages, 7 plates. Mainz: V. Zabern.

Chabanaud, P.

1931. Notes ichthyologiques, II: Sur la présence de côtes crâniennes chez un *Gonorynchus* Gronov. *Bulletin de la Société Zoologique de France,* 56:115–118.

1942. Les côtes et le squelette intermusculaire de *Clupea harengus. Comptes Rendus des Séances de l'Académie des Sciences, Paris,* 215:380–382.

Chang, M.-M., and C.-C. Chou

1976. Discovery of *Plesiolycoptera* in Songhuajiang-Liaoning Basin and Origin of Osteoglossomorpha. *Vertebrata Palasiatica,* 14:146–153. [In Chinese.]

1977. On Late Mesozoic Fossil Fishes from Zhejiang Province, China. *Memoirs of the Institute of Vertebrate Paleontology and Paleoanthropology, Academia Sinica,* 12:1–59.

Chapleau, F.

1993. Pleuronectiform Relationships: A Cladistic Reassessment. *Bulletin of Marine Science,* 52:516–540.

Chapman, W.M.

1942. The Osteology and Relationships of the Argentinidae, a Family of Oceanic Fishes. *Journal of the Washington Academy of Sciences,* 32:104–117.

1944. The Osteology of the Pacific Deep-bodied Anchovy, *Anchoa compressa. Journal of Morphology,* 74:311–329.

Collette, B.B., and L.N. Chao

1975. Systematics and Morphology of the Bonitos (*Sarda*) and Their Relatives (Scombridae, Sardini). *Fishery Bulletin,* 73:516–625.

Daget, J., and F. D'Aubenton

1957. Développement et morphologie du crâne d'*Heterotis niloticus* Ehr. *Bulletin de l'Institut Francais d'Afrique Noire,* A19:881–936.

Dingerkus, G., and L.D. Uhler

1977. Enzyme Clearing of Alcian Blue Stained Whole Small Vertebrates for Demonstration of Cartilage. *Stain Technology,* 52:229–232.

Emelianov, S.W.

1935. Die Morphologie der Fischrippen. *Zoologische Jahrbücher, Anatomie,* 60:132–262.

Farris, J.S.

1988. Hennig86 Version 1.5. Port Jefferson, New York: published by the author.

Faruqi, A.J.

1935. The Development of the Vertebral Column in the Haddock (*Gadus aeglifinus*). *Proceedings of the Zoological Society of London,* 1935:313–332.

Fink, S.V., and W.L. Fink

1981. Interrelationships of the Ostariophysan Fishes. *Zoological Journal of the Linnean Society,* 72:297–353.

Fink, W.L.

1984. Basal Euteleosts: Relationships. *Special Publications, American Society of Ichthyologists and Herpetologists,* 1:202–206.

Fink, W.L., and S.H. Weitzman

1982. Relationships of the Stomiiform Fishes (Teleostei), with a Description of *Diplophos. Bulletin of the Museum of Comparative Zoology,* 150:31–93.

Forey, P.L.

1973. A Revision of the Elopiform Fishes, Fossil and Recent. *Bulletin of the British Museum (Natural History), Geology,* supplement, 10:1–222.

Fraser, T.H.

1968. Comparative Osteology of the Atlantic Snooks (Pisces, *Centropomus*). *Copeia,* 1968:433–460.

Fujita, K.

1990. *The Caudal Skeleton of Teleostean Fishes.* xiii + 897 pages. Tokyo: Tokai University Press.

Gardiner, B.G.

1984. The Relationships of the Palaeoniscid Fishes, a Review Based on New Specimens of *Mimia* and *Moythomasia* from the Upper Devonian of Western Australia. *Bulletin of the British Museum (Natural History), Geology,* 36:173–428.

Gayet, M., and M. Chardon

1987. Possible Otophysic Connections in Some Fossil and Living Ostariophysan Fishes. *In* S.O. Kullander and B. Fernholm, editors, *Proceedings of the Fifth Congress of European Ichthyologists,* pages 31–42. Stockholm: Swedish Museum of Natural History.

Goodrich, E.S.

1909. Vertebrata Craniata, First Fascicle: Cyclostomes and Fishes. *In* R. Lankester, editor, *A Treatise on Zoology,* 9:xvii + 518 pages. London: A. and C. Black.

1930. *Studies on the Structure and Development of Vertebrates.* xxx + 837 pages. London: MacMillan.

Göppert, E.
1895. Untersuchungen zur Morphologie der Fischrippen. *Morphologisches Jahrbuch*, 23:145–217.

Gosline, W.A.
1960. Contributions toward a Classification of Modern Isospondylous Fishes. *Bulletin of the British Museum (Natural History), Zoology*, 6:325–365.
1969. The Morphology and Systematic Position of the Alepocephaloid Fishes. *Bulletin of the British Museum (Natural History), Zoology*, 18:183–218.
1970. A Reinterpretation of the Teleostean Fish Order Gobiesociformes. *Proceedings of the California Academy of Sciences*, 38:363–382.

Grande, L.
1985. Recent and Fossil Clupeomorph Fishes, with Materials for Revision of the Subgroups of Clupeoids. *Bulletin of the American Museum of Natural History*, 181:231–372.

Grande, L., and W.E. Bemis
1991. Osteology and Phylogenetic Relationships of Fossil and Recent Paddlefishes (Polyodontidae) with Comments on the Interrelationships of Acipenseriformes. *Journal of Vertebrate Paleontology*, 11, supplement, 1:vii + 1–121.

Grande, L., and G. Nelson
1985. Interrelationships of Fossil and Recent Anchovies (Teleostei: Engrauloidea) and a Description of a New Species from the Miocene of Cyprus. *American Museum Novitates*, 2826:1–16.

Greenwood, P.H.
1968. The Osteology and Relationships of the Denticipitidae, a Family of Clupeomorph Fishes. *Bulletin of the British Museum (Natural History), Zoology*, 16:213–273.
1973. Interrelationships of Osteoglossomorphs. *In* P.H. Greenwood, R.S. Miles, and C. Patterson, editors, *Interrelationships of Fishes*, pages 307–332. London: Academic Press.
1976. A Review of the Family Centropomidae (Pisces, Perciformes). *Bulletin of the British Museum (Natural History), Zoology*, 29:1–81.

Greenwood, P.H., and D.E. Rosen
1971. Notes on the Structure and Relationships of the Alepocephaloid Fishes. *American Museum Novitates*, 2473:1–41.

Greenwood, P.H., D.E. Rosen, S.H. Weitzman, and G.S. Myers
1966. Phyletic Studies of Teleostean Fishes, with a Provisional Classification of Living Forms. *Bulletin of the American Museum of Natural History*, 131:339–456.

Günther, A.C.L.G.
1859. *Catalogue of the Acanthopterygian Fishes in the Collection of the British Museum*. Volume 1, xxxii + 524 pages. London: British Museum.

Hartel, K.E., and M.L.J. Stiassny
1986. The Identification of Larval *Parasudis* (Teleostei, Chlorophthalmidae); with Notes on the Anatomy and Relationships of the Aulopiform Fishes. *Breviora*, 487:1–23.

Hensley, D.A.
1977. Larval Development of *Engyophrys senta* (Bothidae), with Comments on Intermuscular Bones in Flatfishes. *Bulletin of Marine Science*, 27:681–703.

Hensley, D.A., and E.H. Ahlstrom
1984. Pleuronectiformes: Relationships. *Special Publications, American Society of Ichthyologists and Herpetologists*, 1:670–687.

Howes, G.J.
1985. Cranial Muscles of Gonorynchiform Fishes, with Comments on Generic Relationships. *Bulletin of the British Museum (Natural History), Zoology*, 49:273–303.

Jamieson, B.G.M.
1991. *Fish Evolution and Systematics: Evidence from Spermatozoa*. xiv + 319 pages. Cambridge, England: Cambridge University Press.

Johnson, G.D.
1984. Percoidei: Development and Relationships. *In* H.G. Moser, W.J. Richards, D.M. Cohen, M.P. Fahay, A.W. Kendall, Jr., and S.L. Richardson, editors, *Ontogeny and Systematics of Fishes*, pages 464–498. American Society of Ichthyologists and Herpetologists, Special Publications, 1.
1986. Scombroid Phylogeny: An Alternative Hypothesis. *Bulletin of Marine Science*, 39:1–41.
1992. Monophyly of the Euteleostean Clades—Neoteleostei, Eurypterygii, and Ctenosquamata. *Copeia*, 1992:8–25.

Johnson, G.D., and E.B. Brothers
1993. *Schindleria:* A Paedomorphic Goby. *Bulletin of Marine Science*, 52:441–471.

Johnson, G.D., and A.C. Gill
1994. Perches and Allies. In *Encyclopedia of Animals: Fishes*, pages 181–196. Sydney: University of New South Wales Press.

Johnson, G.D., M. Okiyama, and Y. Tominaga
1989. Osteology and Relationships of *Pseudotrichonotus altivelis*. *Abstracts of Third Indo-Pacific Fish Conference*, page 49.

Johnson, G.D., and C. Patterson
1993. Percomorph Phylogeny: A Survey and a New Proposal. *Bulletin of Marine Science*, 52:554–626.

Johnson, R.K.
1982. Fishes of the Families Evermanellidae and Scopelarchidae: Systematics, Morphology, Interrelationships, and Zoogeography. *Fieldiana, Zoology*, new series, 12:xiii + 1–252.
1984. Giganturidae: Development and Relationships. *Special Publications, American Society of Ichthyologists and Herpetologists*, 1:199–201.

Johnson, R.K., and E. Bertelsen
1991. The Fishes of the Family Giganturidae: Systematics, Development, Distribution and Aspects of Biology. *Dana-Report*, 91:1–45.

Jollie, M.
1962. *Chordate Morphology*. xiv + 478 pages. New York: Reinhold.
1975. Development of the Head Skeleton and Pectoral Girdle in *Esox*. *Journal of Morphology*, 147:61–88.
1984a. Development of the Head and Pectoral Skeleton of *Amia* with a Note on the Scales. *Gegenbaurs Morphologisches Jahrbuch*, 130:315–351.
1984b. Development of the Head and Pectoral Skeleton of *Polypterus* with a Note on Scales (Pisces: Actinopterygii). *Journal of Zoology*, 204:469–507.

Jungersen, H.F.E.
1910. Ichthyotomical Contributions, II: The Structure of the Aulostomidae, Syngnathidae and Solenostomidae. *Kongelige Danske Videnskabernes, Selskabs Skrifter*, 7(8):267–364.

Kafuku, T.
1950. "Red Muscles" in Fishes, I: Comparative Anatomy of the Scombroid Fishes of Japan. *Japanese Journal of Ichthyology*, 1:89–100. [In Japanese, with English summary.]

Kishinouye, K.
1923. Contributions to the Comparative Anatomy of the So-called Scombroid Fishes. *Journal of the College of Agriculture, Imperial University of Tokyo*, 8:293–475, plates 13–34.

Lauder, G.V., and K.F. Liem
1983. The Evolution and Interrelationships of the Actinopterygian Fishes. *Bulletin of the Museum of Comparative Zoology*, 150:95–197.

Le, H.L.V., G. Lecointre, and R. Perasso
1993. A 28S rRNA-based Phylogeny of the Gnathostomes: First Steps in the Analysis of Conflict and Congruence with Morphologically Based Cladograms. *Molecular Phylogenetics and Evolution*, 2:31–51.

Lenglet, G.
1974. Contribution a l'étude ostéologique des Kneriidae. *Annales de la Société Royale Zoologique de Belgique*, 104:51–103.

Leviton, A.E, R.H. Gibbs, Jr., E. Heal, and C.E. Dawson
1985. Standards in Herpetology and Ichthyology, Part I: Standard Symbolic Codes for Institutional Resource Collections in Herpetology and Ichthyology. *Copeia,* 1985:802–832.
Li, G.
1987. A New Genus of Hiodontidae from Luozigou Basin, East Jilin. *Vertebrata Palasiatica,* 25:91–107.
Ma, F.
1983. Early Cretaceous Primitive Teleosts from the Jiaohe Basin of Jilin Province, China. *Vertebrata Palasiatica,* 21:17–31.
1986. On the Generic Status of *Lycoptera tungi. Vertebrata Palasiatica,* 24:260–268.
Mabee, P.M.
1988. Supraneural and Predorsal Bones in Fishes: Development and Homologies. *Copeia,* 1988:827–838.
1993. Phylogenetic Interpretation of Ontogenetic Change: Sorting Out the Actual and Artefactual in an Empirical Case Study of Centrarchid Fishes. *Zoological Journal of the Linnean Society,* 107:175–291.
Mago Leccia, F.
1978. Los peces de la familia Sternopygidae de Venezuela. *Acta Scientifica Venezolana,* 29, supplement, 1:1–89.
Markle, D.F.
1989. Aspects of Character Homology and Phylogeny of the Gadiformes. *Science Series, Natural History Museum of Los Angeles County,* 32:59–88.
McDowall, R.M.
1969. Relationships of Galaxioid Fishes with a Further Discussion of Salmoniform Classification. *Copeia,* 1969:796–824.
1971. Fishes of the Family Aplochitonidae. *Journal of the Royal Society of New Zealand,* 1:31–52.
1976. Fishes of the Family Prototroctidae (Salmoniformes). *Australian Journal of Marine and Freshwater Research,* 27:641–659.
1978. A Revision of the Genus *Paragalaxias* Scott (Salmoniformes: Galaxiidae). *Australian Journal of Marine and Freshwater Research,* 29:93–108.
Monod, T.
1960. A propos du *pseudobrachium* des *Antennarius* (Pisces, Lophiiformes). *Bulletin de l'Institut Francais d'Afrique Noire,* A22:620–698.
1963. Sur quelques points de l'anatomie de *Gonorhynchus gonorhynchus* (Linné 1766). *Mémoires de l'Institut Francais d'Afrique Noire,* 68:255–313.
1968. Le complexe urophore des poissons téléostéens. *Mémoires de l'Institut Fondamental d'Afrique Noire,* 81:i–vi + 1–705.
Moore, J.A.
1993. The Phylogeny of the Trachichthyiformes (Teleostei: Percomorpha). *Bulletin of Marine Science,* 52:114–136.
Müller-Schmid, A., B. Ganss, T. Gorr, and W. Hoffmann
1993. Molecular Analysis of Ependymins from the Cerebrospinal Fluid of the Orders Clupeiformes and Salmoniformes: No Indication for the Existence of an Euteleost Infradivision. *Journal of Molecular Evolution,* 36:578–585.
Nelson, G.J.
1969. Infraorbital Bones and Their Bearing on the Phylogeny and Geography of Osteoglossomorph Fishes. *American Museum Novitates,* 2394:1–37.
1970a. The Hyobranchial Apparatus of Teleostean Fishes of the Families Engraulidae and Chirocentridae. *American Museum Novitates,* 2410:1–30.
1970b. Gill Arches of Some Teleostean Fishes of the Families Salangidae and Argentinidae. *Japanese Journal of Ichthyology,* 17:61–66.
1972. Cephalic Sensory Canals, Pitlines, and the Classification of Esocoid Fishes, with Notes on Galaxiids and Other Teleosts. *American Museum Novitates,* 2492:1–49.
1973. Relationships of Clupeomorphs, with Remarks on the Structure of the Lower Jaw in Fishes. *In* P.H. Greenwood, R.S. Miles, and C. Patterson, editors, *Interrelationships of Fishes,* pages 333–349. London: Academic Press.
Nelson, G.J., and P.Y. Ladiges
1992a. TAS and TAX: MSDOS Programs for Cladistics. New York and Melbourne: published by the authors.
1992b. Information Content and Fractional Weight of Three-item Statements. *Systematic Biology,* 41(4):490–494.
Nelson, J.S.
1984. *Fishes of the World,* Second edition, xv + 523 pages. New York: Wiley.
Norden, C.R.
1961. Comparative Osteology of Representative Salmonid Fishes, with Particular Reference to the Grayling (*Thymallus arcticus*) and Its Phylogeny. *Journal of the Fisheries Research Board of Canada,* 18:679–791.
Nursall, J.R.
1956. The Lateral Musculature and the Swimming of Fish. *Proceedings of the Zoological Society of London,* 126:127–143.
Okamura, O., and H. Kishimoto
1993. *Selenoscopus turbisquamatus,* a New Genus and Species of Uranoscopid Fish from Japan and the Norfolk Ridge. *Japanese Journal of Ichthyology,* 39:311–317.
Okiyama, M.
1984. Myctophiformes: Relationships. *Special Publications, American Society of Ichthyologists and Herpetologists,* 1:254–259.
Olney, J.E., G.D. Johnson, and C.C. Baldwin
1993. Phylogeny of Lampridiform Fishes. *Bulletin of Marine Science,* 52:137–169.
Ørvig, T.
1962. Y a-t-il une relation directe entre les arthrodires ptyctodontides et les holocéphales? *Colloques Internationaux du Centre National de la Recherche Scientifique,* 104:49–61.
Owen, R.
1846. *Lectures on the Comparative Anatomy and Physiology of the Vertebrate Animals, Part 1: Fishes.* xi + 308 pages. London: Longman, Brown, Green and Longmans.
1866. *The Anatomy of Vertebrates, Volume 1: Fishes and reptiles.* xlii + 600 pages. London: Longmans, Green and Co.
Patterson, C.
1964. A Review of Mesozoic Acanthopterygian Fishes, with Special Reference to Those of the English Chalk. *Philosophical Transactions of the Royal Society of London,* series B, 247:213–482.
1970. Two Upper Cretaceous Salmoniform Fishes from the Lebanon. *Bulletin of the British Museum (Natural History), Geology,* 19:205–296.
1975. The Braincase of Pholidophorid and Leptolepid Fishes, with a Review of the Actinopterygian Braincase. *Philosophical Transactions of the Royal Society of London,* B269:275–579.
1977. Cartilage Bones, Dermal Bones and Membrane Bones, or the Exoskeleton Versus the Endoskeleton. *In* S.M. Andrews, R.S. Miles, and A.D. Walker, editors, *Problems in Vertebrate Evolution,* pages 77–121. London: Academic Press.
1984. *Chanoides,* a Marine Eocene Otophysan Fish (Teleostei: Ostariophysi). *Journal of Vertebrate Paleontology,* 4:430–456.
1993. Osteichthyes: Teleostei. *In* M.J. Benton, editor, *The Fossil Record 2,* pages 621–656. London: Chapman and Hall.
Patterson, C., and D.E. Rosen
1977. Review of Ichthyodectiform and Other Mesozoic Teleost Fishes and the Theory and Practice of Classifying Fossils. *Bulletin of the American Museum of Natural History,* 158:81–172.
1989. The Paracanthopterygii Revisited: Order and Disorder. *Science Series, Natural History Museum of Los Angeles County,* 32:5–36.

Phillips, J.B.
1942. Osteology of the Sardine (*Sardinops caerulea*). *Journal of Morphology*, 70:463–500.
Pietsch, T.W.
1989. Phylogenetic Relationships of Trachinoid Fishes of the Family Uranoscopidae. *Copeia*, 1989:253–303.
Pietsch, T.W., and C.P. Zabetian
1990. Osteology and Interrelationships of the Sand Lances (Teleostei: Ammodytidae). *Copeia*, 1990:78–100.
Ramanujam, S.G.M.
1929. The Study of the Development of the Vertebral Column in Teleosts, as Shown in the Life-history of the Herring. *Proceedings of the Zoological Society of London*, 1929:365–414.
Ridewood, W.G.
1904. On the Cranial Osteology of the Fishes of the Families Elopidae and Albulidae, with Remarks on the Morphology of the Skull in the Lower Teleostean Fishes Generally. *Proceedings of the Zoological Society of London*, 1904(2):35–81.
1905. On the Cranial Osteology of the Fishes of the Families Osteoglossidae, Pantodontidae, and Phractolaemidae. *Journal of the Linnean Society, Zoology*, 29:252–282.
Roberts, T.R.
1984. Skeletal Anatomy and Classification of the Neotenic Asian Salmoniform Superfamily Salangoidea (Icefishes or Noodlefishes). *Proceedings of the California Academy of Sciences*, 43(13):179–220.
Rofen, R.R.
1966a. Family Omosudidae. *Memoirs of the Sears Foundation for Marine Research*, 1(5):462–481.
1966b. Family Anotopteridae. *Memoirs of the Sears Foundation for Marine Research*, 1(5):498–510.
1966c. Family Evermanellidae. *Memoirs of the Sears Foundation for Marine Research*, 1(5):511–565.
Rosen, D.E.
1962. Comments on the Relationships of the North American Cave Fishes of the Family Amblyopsidae. *American Museum Novitates*, 2109: 1–35.
1964. The Relationships and Taxonomic Position of the Halfbeaks, Killifishes, Silversides, and Their Relatives. *Bulletin of the American Museum of Natural History*, 127:217–268.
1973. Interrelationships of Higher Euteleostean Fishes. *In* P.H. Greenwood, R.S. Miles, and C. Patterson, editors, *Interrelationships of Fishes*, pages 397–513. London: Academic Press.
1974. Phylogeny and Zoogeography of Salmoniform Fishes and Relationships of *Lepidogalaxias salamandroides*. *Bulletin of the American Museum of Natural History*, 153:265–326.
1985. An Essay on Euteleostean Classification. *American Museum Novitates*, 2827:1–57.
Rosen, D.E., P.L. Forey, B.G. Gardiner, and C. Patterson
1981. Lungfishes, Tetrapods, Paleontology, and Plesiomorphy. *Bulletin of the American Museum of Natural History*, 167:159–276.
Rosen, D.E., and P.H. Greenwood
1970. Origins of the Weberian Apparatus and the Relationships of the Ostariophysan and Gonorynchiform Fishes. *American Museum Novitates*, 2428:1–25.
Saether, O.A.
1983. The Canalized Evolutionary Potential: Inconsistencies in Phylogenetic Reasoning. *Systematic Zoology*, 32:343–359.
Sanford, C.P.J.
1988. The Phylogenetic Relationships of the Salmonoid Fishes. 361 pages. Doctoral dissertation, University of London, London.
1990. The Phylogenetic Relationships of Salmonoid Fishes. *Bulletin of the British Museum (Natural History), Zoology*, 56:145–153.
Sawada, Y.
1982. Phylogeny and Zoogeography of the Superfamily Cobitoidea (Cyprinoidei, Cypriniformes). *Memoirs of the Faculty of Fisheries, Hokkaido University*, 28:65–223.
Schaeffer, B., and C. Patterson
1984. Jurassic Fishes from the Western United States, with Comments on Jurassic Fish Distribution. *American Museum Novitates*, 2796:1–86.
Siebenrock, F.
1901. Ueber die Verbindungsweise des Schultergürtels mit dem Schädel bei den Teleosteern, Eine morphologische Studie. *Annalen des Kaiserlich-Köngliches Naturhistorischen Hofmuseums, Wien*, 16:105–141.
Smith, D.G.
1989. Family Congridae. *Memoirs of the Sears Foundation for Marine Research*, 1(9):460–567.
Springer, V.G.
1988. *Rotuma lewisi*, New Genus and Species of Fish from the Southwest Pacific (Gobioidei, Xenisthmidae). *Proceedings of the Biological Society of Washington*, 101(3):530–539.
Starks, E.C.
1904a. The Osteology of Some Berycoid Fishes. *Proceedings of the United States National Museum*, 27:601–619.
1904b. A Synopsis of Characters of Some Fishes Belonging to the Order Haplomi. *Biological Bulletin of the Marine Biological Station, Woods Hole, Massachusetts*, 7:254–262.
1923. The Osteology and Relationships of the Uranoscopoid Fishes. *Stanford University Publications, University Series, Biological Sciences*, 3(3):255–290.
Stewart, J.D.
1984. Taxonomy, Paleoecology, and Stratigraphy of the Halecostome-Inoceramid Associations of the North American Upper Cretaceous Epicontinental Seaways. 201 pages. Doctoral dissertation, University of Kansas, Lawrence, Kansas.
Stiassny, M.L.J.
1986. The Limits and Relationships of the Acanthomorph Teleosts. *Journal of Zoology, London*, series B, 1:411–460.
1990. Notes on the Anatomy and Relationships of the Bedotiid Fishes of Madagascar, with a Taxonomic Revision of the Genus *Rheocles* (Atherinomorpha: Bedotiidae). *American Museum Novitates*, 2979:1–27.
Stiassny, M.L.J., and J.A. Moore
1992. A Review of the Pelvic Girdle of Acanthomorph Fishes, with Comments on Hypotheses of Acanthomorph Intrarelationships. *Zoological Journal of the Linnean Society*, 104:209–242.
Swofford, D.L.
1990. PAUP: Phylogenetic Analysis Using Parsimony, Version 3.0. Champaign, Ilinois: Illinois Natural History Survey.
Taverne, L.
1972. Ostéologie des genres *Mormyrus* Linné, *Mormyrops* Müller, *Hyperopisus* Gill, *Isichthys* Gill, *Myomyrus* Boulenger, *Stomatorhinus* Boulenger et *Gymnarchus* Cuvier; Considérations générales sur la systématique des poissons de l'ordre des Mormyriformes. *Annales, Musée Royal de l'Afrique Centrale*, Series 8, Sciences Zoologiques, 200:1–194.
1974. L'ostéologie d'*Elops* Linné, C., 1766 (Pisces Elopiformes) et son intéret phylogénétique. *Mémoires, Académie Royale de Belgique, Classe des Sciences*, 8 volumes, 2(41)2:1–96.
1977a. Ostéologie et position systématique du genre *Thrissops* Agassiz, 1833 (*sensu stricto*) (Jurassique supérieur de l'Europe occidentale) au sein des téléostéens primitifs. *Geobios*, 10:5–33.
1977b. Ostéologie, phylogénèse et systématique des téléostéens fossiles et actuels du super-ordre des ostéoglossomorphes, Première partie: Ostéologie des genres *Hiodon*, *Eohiodon*, *Lycoptera*, *Osteoglossum*,

Scleropages, Heterotis et *Arapaima*. *Mémoires, Académie Royale de Belgique, Classe des Sciences,* 8 volumes, 2(42)3:1–235.

1978. Ostéologie, phylogénèse et systématique des téléostéens fossiles et actuels du super-ordre des ostéoglossomorphes, Deuxième partie: Ostéologie des genres *Phareodus, Phareoides, Brychaetus, Musperia, Pantodon, Singida, Notopterus, Xenomystus* et *Papyrocranus*. *Mémoires, Académie Royale de Belgique, Classe des Sciences,* 8 volumes 2(42)6:1–213.

Thys van den Audenaerde, D.F.E.

1961. L'anatomie de *Phractolaemus ansorgei* Blgr. et la position systématique des Phractolaemidae. *Annales, Musée Royal de l'Afrique Centrale,* Series 8, Sciences Zoologiques, 103:99–167.

Tyler, J.C., G.D. Johnson, I. Nakamura, and B.B. Collette

1989. Morphology of *Luvarus imperialis* (Luvaridae), with a Phylogenetic Analysis of the Acanthuroidei (Pisces). *Smithsonian Contributions to Zoology,* 485:1–78.

Weitzman, S.H.

1962. The Osteology of *Brycon meeki,* a Generalized Characid Fish, with an Osteological Definition of the Family. *Stanford Ichthyological Bulletin,* 8:1–77.

1974. Osteology and Evolutionary Relationships of the Sternoptychidae, with a New Classification of Stomiatoid Families. *Bulletin of the American Museum of Natural History,* 153:327–478.

Westneat, M.W., W. Hoese, C.A. Pell, and S.A. Wainwright

1993. The Horizontal Septum: Mechanisms of Force Transfer in Locomotion of Scombrid Fishes (Scombridae, Perciformes). *Journal of Morphology,* 217:183–204.

Whitehead, P.J.P., and G.G. Teugels

1985. The West African Pygmy Herring *Sierrathrissa leonensis:* General Features, Visceral Anatomy, and Osteology. *American Museum Novitates,* 2835:1–44.

Wilson, M.V.H.

1984. Osteology of the Palaeocene Teleost *Esox tiemani*. *Palaeontology,* 27:597–608.

1992. Importance for Phylogeny of Single and Multiple Stem-group Fossil Species with Examples from Freshwater Fishes. *Systematic Biology,* 41:462–470.

Wilson, M.V.H., and P. Veilleux

1982. Comparative Osteology and Relationships of the Umbridae (Pisces: Salmoniformes). *Zoological Journal of the Linnean Society,* 76: 321–352.

Winterbottom, R.

1993. The Search for the Gobioid Sister Group (Actinopterygii; Percomorpha). *Bulletin of Marine Science,* 52:395–414.

Yabumoto, Y., and T. Uyeno

1982. Osteology of the Clupeiform Fish, Genus *Hyperlophus* (II). *Bulletin, Kitakyushu Museum of Natural History,* 4:77–102.

Zehren, S.J.

1979. The Comparative Osteology and Phylogeny of the Beryciformes. *Evolutionary Monographs,* 1:1–389.

Plates

PLATE 1

A–C, *Albula vulpes* (L.), USNM 128509, 51 mm SL; D, *Salmo trutta* L., BMNH 1981.9.22.81, 52 mm SL; E, F, *Myctophum obtusirostre* Tåning, AMNH 29140 SW, 40 mm SL; G, *Cynothrissa mento* Regan, BMNH 1967.12.29.1, 100 mm SL; H, *Galaxias fasciatus* Gray, BMNH 1965.12.6.15, 84 mm SL. A shows the posterior abdominal region in *Albula,* with the epipleural series of ligaments (attaching to ribs) containing epipleural bones at right of picture but without bone at left. Epicentral series of ligaments (attaching to parapophysis in this region) are visible just below middle of vertebral centra. B and C show the epicentral series of *Albula* in close-up, with B in dorsolateral view to show ligaments against ribs, and C in lateral view to show them against centra, with epipleural ligaments attaching to ribs below them. E and F show the three series of intermusculars in *Myctophum.* E is in the caudal region, with faintly stained, proximally forked epineural and epipleural bones respectively above and below vertebral centra. Epicentral ligaments are shown against vertebral centra by incident light, and they are forked distally in this region. F is in the abdominal region, and it shows epineural bones attaching to neural arches and epicentral ligaments attaching to ribs or parapophyses. G is in the posterior abdominal region of the pellonuline clupeoid *Cynothrissa,* which shows cartilaginous chevrons at tips of epicentral bones. There are proximally forked epineural and epipleural bones respectively above and below the vertebral centra. D and H show cartilage rods in the distal part of epicentral ligaments in the abdominal region of a salmonoid (D) and an osmeroid (H). Both are in dorsolateral view, with epineurals seen as overlying vertebral centra, and epicentrals seen as overlying ribs. In H, the epineurals and ribs are unstained, although ossified.

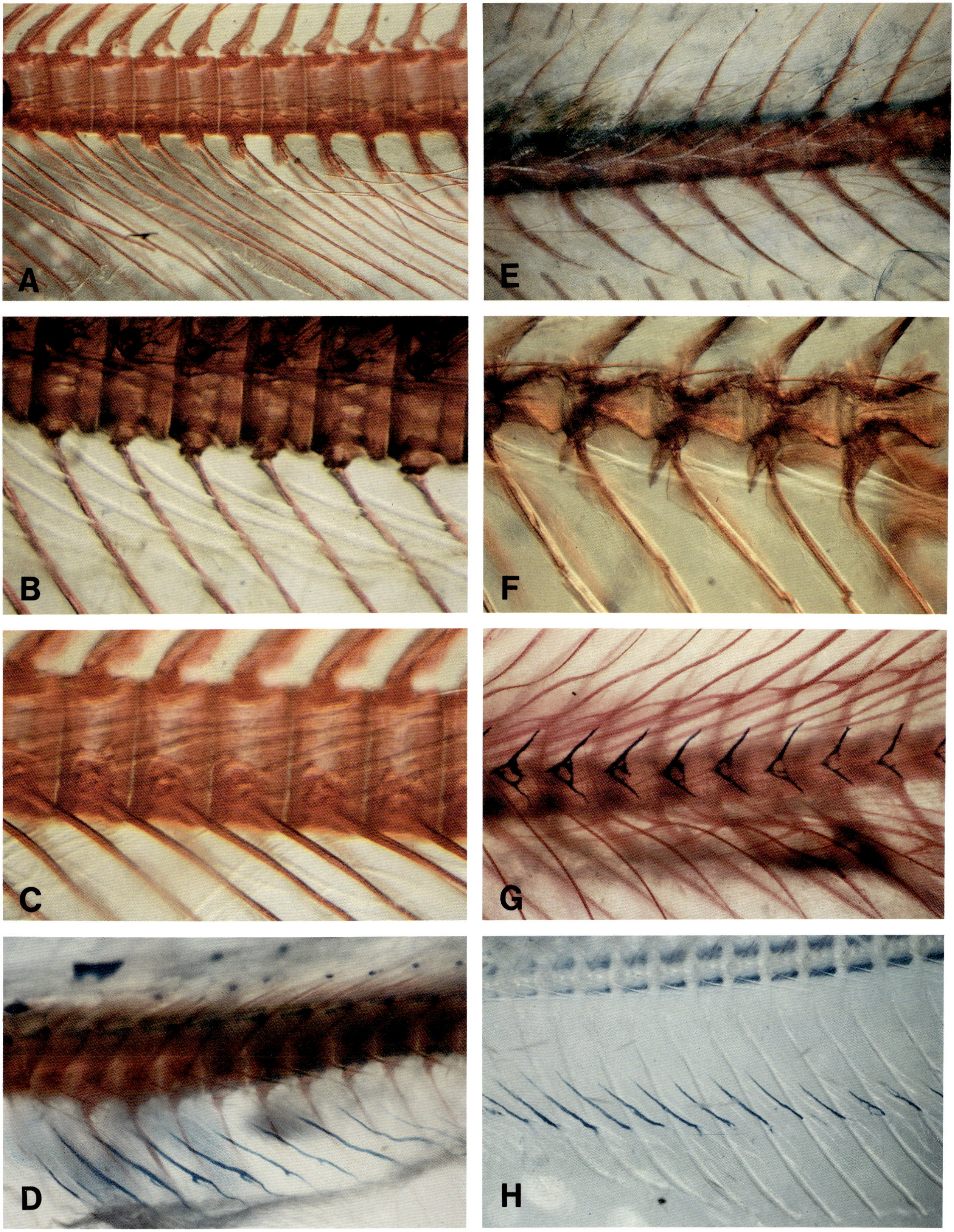
A
E
B
F
C
G
D
H

PLATE 2

A–C, G, H, *Polymixia lowei* Günther, USNM 308378, 72 mm SL; D, *Polymixia lowei* Günther, BMNH 1987.12.7.1, 128 mm SL; E, F, *Myripristis murdjan* (Forskål), AMNH 43444SW, 51 mm SL. A–C show the posterior abdominal and anterior caudal region in *Polymixia* under different lighting conditions. The epineural bones attach to the parapophysis at left of photographs and move up to the neural spines at right. The epipleural bones are at the bottom right, with the ligamentous attachment of the first one most clearly seen in B. The epicentral series of ligaments, attaching to rib (at left), parapophysis, or haemal arch, is invisible in A, visible in transmitted light in B, and emphasized by incident light in C. D shows the distal cartilaginous rods in anterior epicentral ligaments of a larger *Polymixia*. The epineural bones are at top of photograph and the lateral line nerve is at the tip of the epicentrals. E and F show the epicentral ligaments on the last few abdominal and first few caudal vertebrae in *Myripristis*, in transmitted light, with a close-up in F. The posterior epineural bones, visible in E, attach to the neural arches and spines. G and H show the posterior members of the epipleural series in *Polymixia*, with transition from a ligament containing bone distally to a ligament without bone in G, and close-up of a ligament with enclosed bone in H.

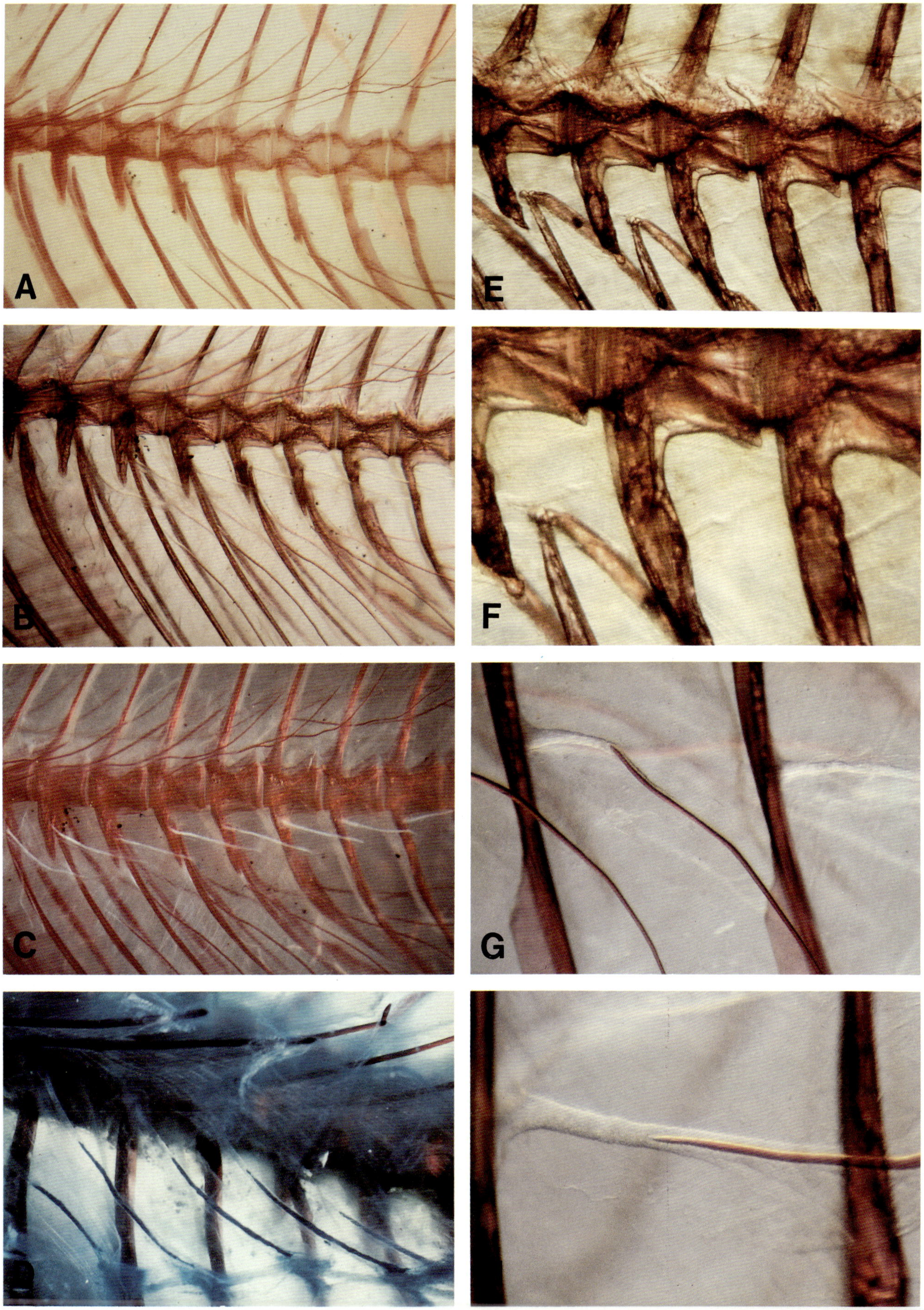
A
B
C
E
F
G

Table 3.---Distribution of intermuscular bones and ligaments in certain osteoglossomorph, elopomorph and clupeomorph teleosts. Numbers in top horizontal row = vertebrae, numbered from the first.

		Occ	ANA	1	2	3	4	5	6	7	8	9	10	11	12	13	14	15	16	17	18	19	20	21	22	23	24	25	26	27	28	29	30
Hiodon	En	L	-	F_1	F_1	F_1	F_1	F_1	F_1	F_1	F_1	F_1	F_1	F_1	F_1	F_1	F_1	F_1	F_1	F_1	F_1	F_1	F_1	F_1	F_1	F_1	F_1	F_1	F_1	F_1	F_1	F_1	L_1
alosoides	Ec	L		L_2	L_3	L_3	L_3	L_3	L_3	L_3	L_3	L_3	L_3	L_3	L_3	L_3	L_3	L_3	L_3	L_3	L_3	L_3	L_3	L_3	L_3	L_3	L_3	L_3	L_3	L_3	L_3	L_3	L_3
AMNH 23754SW	Ep																													L_4	L_4	L_4	L_4
SL 55 mm	R			S		B	B	B	B	B	B	B	B	B	B	B	B	B	B	B	B	B	B	B	B	B	B	B	B	B	B	B	B
Osteoglossum	En	BB	-	F_1	F_1	F_1	F_1	F_1	F_1	F_1	F_1	F_1	F_1	F_1	F_1	B_1	B_1	B_1	B_1	B_1	B_1	B_1	B_1	B_1	B_1	B_1	B_1	B_1	B_1	B_1	B_1	B_1	B_1
bicirrhosum	Ec			L_2	L_3	L_3	L_3	L_3	L_3	L_3	L_3	L_3	L_3	L_3	L_3	L_3	L_3	L_3	L_3	L_3	L_3	L_3	L_3	L_3	L_3	L_3	L_3	L_3	L_3	L_3	L_2	L_2	L_2
USNM 198123	Ep																										L_4	L_4	L_4	L_4	L_3	L_5	L_5
SL 59 mm	R	S		S	B	B	B	B	B	B	B	B	B	B	B	B	B	B	B	B	B	B	B	B	B	B	B	B	B	B			
Heterotis	En		-	B_1	B_1	B_1	B_1	B_1	B_1	B_1	B_1	B_1	B_1	B_1	B_1	B_1	B_1	B_1	B_1	B_1	B_1	B_1	B_1	B_1	B_1	B_1	B_1	B_1	B_1	B_1	B_1	B_1	B_1
niloticus	Ec	———	———		L_2	L_4	L_4	L_4	L_4	L_4	L_4	L_4	L_4	L_4	L_4	L_4	L_4	L_4	L_4	L_4	L_4	L_4	L_4	L_4	L_4	L_3	L_3	L_3	L_3	L_5	L_5	L_5	L_2
BMNH uncat.	Ep																													B_5	B_5	B_5	B_5
SL 68 mm	R	S		S	B	B	B	B	B	B	B	B	B	B	B	B	B	B	B	B	B	B	B	B	B	B	B	B	B				
Xenomystus	En	B	-	B_1	B_1	B_1	B_2	B_2	B_2	B_2	L_2M	L_2M	L_2M	L_2M	L_2M	L_2M	L_2M	L_2M	L_2M	L_2M	L_2M	M	M	M	M	M	M	M	M	M	M	M	M
nigri	Ec				L_1	L_2	L_2	L_2	L_2	L_2	L_2	L_2	L_2	L_2	L_2	L_2	L_2	L_2	L_2	L_2	L_2	L_2	L_2	L_2	L_2	L_2	L_2	L_2	L_2	L_2	L_2	L_2	L_2
BMNH uncat.	Ep						M	M	M	M	M	L_3M	L_3M	L_5M	L_5M	L_5M	L_5M	L_5M	L_5M	L_5M	L_5M	L_5M	L_5M	L_5M	L_5M	M	M	M	M	M	M	M	M
SL 110 mm	R			S		B	B	B	B																								
Elops	En	B	+	B_1	F_1	F_1	F_1	F_1	F_1	F_1	F_1	F_1	F_1	F_1	F_1	F_1	F_1	F_1	F_1	F_1	F_1	F_1	F_1	F_1	F_1	F_1	F_1	F_1	F_1	F_1	F_1	F_1	vB_1
machnata	Ec	L		L_2	L_2	L_2	L_3	L_3	L_3	L_3	L_3	L_3	L_3	L_3	L_3	L_3	L_3	L_3	L_3	L_3	L_3	L_3	L_3	L_3	L_3	L_3	L_3	L_3	L_3	L_3	L_3	L_3	L_3
BMNH 62.8.28.1	Ep																															B_0	B_0
SL 65 mm	R			S					B	B	B	B	B	B	B	B	B	B	B	B	B	B	B	B	B	B	B	B	B	B	B	B	B
Megalops	En	B	-	F_1	F_1	F_1	F_1	F_1	F_1	F_1	F_1	F_1	F_1	F_1	F_1	F_1	F_1	F_1	F_1	F_1	F_1	F_1	F_1	F_1	F_1	F_1	F_1	F_1	F_1	F_1	F_1	F_1	vB_1
cyprinoides	Ec	B		B_2	B_2	B_2	B_2	B_3	B_3	B_3	B_3	B_3	B_3	B_3	B_3	B_3	B_3	B_3	B_3	B_3	B_3	B_3	B_3	B_3	B_3	B_3	B_3	B_3	B_3	B_3	B_3	B_3	B_2
BMNH 55.9.19.832	Ep																							L_4	L_4	L_4	B_4	B_4	B_4	B_4	B_4
SL 99 mm	R			S			B	B	B	B	B	B	B	B	B	B	B	B	B	B	B	B	B	B	B	B	B	B	B	B	B	B	B
Albula	En	>B>BB	-	B_1	B_1	B_1	B_1	B_1	B_1	B_1	B_1	B_1	B_1	B_1	B_1	B_1	B_1	B_1	B_1	B_1	B_1	B_1	B_1	B_1	B_1	B_1	B_1	B_1	B_1	B_1	B_1	B_1	B_1
vulpes	Ec			L_2	L_2	L_2	L_2	L_3	L_3	L_3	L_4	L_4	L_4	L_4	L_4	L_4	L_4	L_4	L_4	L_4	L_4	L_4	L_4	L_4	L_4	L_4	L_4	L_4	L_4	L_4	L_4	L_4	L_4
BMNH 49.11.29.1	Ep												L_4	L_4	L_4	L_4	L_4	L_4	L_4	L_4	L_4	L_4	L_4	L_4	L_4	L_4	L_4	L_4	L_4	L_4	L_4	L_4	L_4
SL 164 mm	R			S		B	B	B	B	B	B	B	B	B	B	B	B	B	B	B	B	B	B	B	B	B	B	B	B	B	B	B	B
Albula	En	BB	-	B_1	B_1	B_1	B_1	B_1	B_1	B_1	B_1	B_1	B_1	B_1	B_1	B_1	B_1	B_1	B_1	B_1	B_1	B_1	B_1	B_1	B_1	B_1	B_1	B_1	B_1	B_1	B_1	B_1	B_1
vulpes	Ec			L_2	L_2	L_2	L_2	L_2	L_2	L_2	L_3	L_3	L_3	L_4	L_4	L_4	L_4	L_4	L_4	L_4	L_4	L_4	L_4	L_4	L_4	L_4	L_4	L_4	L_3	L_3	L_3	L_3	L_3
USNM 128509	Ep															L_4	L_4	L_4	L_4	L_4	L_4	L_4	L_4	L_4	L_4	L_4	L_4	L_4	L_4	L_4	L_4	L_4	L_4
SL 51 mm	R			S		B	B	B	B	B	B	B	B	B	B	B	B	B	B	B	B	B	B	B	B	B	B	B	B	B	B	B	B
Albula	En	LL	-	L_1	B_1	B_1	B_1	B_1	B_1	B_1	B_1	B_1	B_1	B_1	B_1	B_1	B_1	B_1	B_1	B_1	B_1	B_1	B_1	B_1	B_1	B_1	B_1	B_1	B_1	B_1	B_1	B_1	B_1
vulpes	Ec			L_2	L_2	L_2	L_2	L_2	L_2	L_3	L_3	L_3	L_3	L_3	L_3	L_3	L_3	L_3	L_3	L_3	L_3	L_3	L_3	L_2	L_2	L_2	L_2	L_2	L_2	L_2	L_2	L_2	L_2
USNM 128393	Ep																							L_4	L_4	L_4	L_4	L_4	L_4	L_4	L_4	L_4	L_4
SL 27 mm	R			S		B	B	B	B	B	B	B	B	B	B	B	B	B	B	B	B	B	B	B	B	B	B	B	B	B	B	B	B
Halosaurus	En		-	B_1	B_1	B_1	B_1	B_1	B_1	B_1	B_1	B_1	B_1	vB_1	vB_1	vB_1	vB_1	vB_1	vB_1	vB_1	vB_1	vB_1	vB_1	vB_1	vB_1	vB_1	vB_1	vB_1	vB_1	vB_1	vB_1	vB_1	vB_1
guentheri	Ec	L		L_3	L_3	L_3	L_3	L_3	L_3	L_3	L_3	L_3	L_3	L_3	L_3	L_3	L_3	L_3	L_3	L_3	L_3	L_3	L_3	L_3	L_3	L_3	L_3	L_3	L_3	L_3	L_3	L_3	L_3
USNM 319535	Ep																																
SL 259 mm	R			S				B	B	B	B	B	B	B	B	B	B	B	B	B	B	B	B	B	B	B	B	B	B	B	B	B	B
Notacanthus	En		-	V_2V_3	V_2	V_2	vB_2	vB_2	vB_1	vB_1	vB_1	vB_1	vB_1	vB_1	vB_1	vB_1	vB_1	vB_1	vB_1	vB_1	vB_1	vB_1	vB_1	vB_1	vB_1	vB_1	vB_1	vB_1	vB_1	vB_1	vB_1	vB_1	vB_1
chemnitzi	Ec			L_3	L_3	L_3	L_3	L_3	L_3	L_3	L_3	L_3	L_3	L_3	L_3	L_3	L_3	L_3	L_3	L_3	L_3	L_3	L_3	L_3	L_3	L_3	L_3	L_3	L_3	L_3	L_3	L_3	L_3
USNM 214339	Ep												L_3	L_3	L_3	L_3	L_3	L_3	L_3	L_3	L_3	L_3	L_3	L_3	L_3	L_3	L_3	L_3	L_3	L_3	L_3	L_3	L_3
SL 230 mm	R																	L	L	L	L	L	L	L	L	L	L	L	L	L	L	L	L
Anguilla	En		-	F_1	F_1	F_1	F_1	F_1	B_1	B_1	B_1	B_1	B_1	B_1	B_1	B_1	B_1	B_1	B_1	B_1	B_1	B_1	B_1	B_1	B_1	B_1	B_1	B_1	B_1	B_1	B_1	B_1	B_1
rostrata	Ec			L_2	L_2	L_2	L_2	L_2	L_3	L_3	L_3	L_3	L_3	L_3	L_3	L_3	L_3	L_3	L_3	L_3	L_3	L_3	L_3	L_3	L_3	L_3	L_3	L_3	L_3	L_3	L_3	L_3	L_3
USNM 190974	Ep									L_3	L_3	L_3	L_3	L_3	L_3	L_3	L_3	L_3	L_3	L_3	L_3	L_3	L_3	L_3	L_3	L_3	L_3	L_3	L_3	L_3	L_3	L_3	L_3
SL 121 mm	R			L	L	L	L	L	L	B	B	B	B	B	B	B	B	B	B	B	B	B	B	B	B	B	B	B	B	B	B	B	B
Denticeps	En		-												L_1	L_1	L_1	L_1	L_1	L_1	L_1	L_1	B_1	B_1	vB_1	vB_1	vB_1	vB_1	vB_1	V_0	V_0	V_0	V_0
clupeoides	Ec				B_3	B_4	B_4	B_4	B_4	B_4	B_4	B_4	B_4	B_4	B_4	B_4	B_3	B_3	B_2	B_2	B_2	B_2	B_2	B_2	L_2	L_2	L_2	L_2					
BMNH 69.4.26.1	Ep																			B_5	DB_5	DB_5	DB_5	DB_5	DB_5	DB_5	DB_5	DB_5	DB_5	D_0	D_0
Sl 34 mm	R			S		B	B	B	B	B	B	B	B	B	B	B	B	B	B														
Clupea	En	B	+	B_1	F_1	F_1	F_1	F_1	F_1	F_1	F_1	F_1	F_1	F_1	F_1	F_1	F_1	F_1	F_1	F_1	F_1	F_1	F_1	F_1	F_1	B_1	B_1	B_1	vB_1	vB_1	vB_1	vB_1	vB_1
harengus	Ec	B		cB_1	cB_2	cF_4	cF_4	cF_4	cF_4	cF_4	cF_4	cF_4	cF_4	cF_4	cF_4	cF_4	cF_4	cF_4	cF_4	F_4	F_4	F_4	F_4	F_4	F_4	F_4	B_3	B_3	B_3	B_2	B_2	B_2	B_2
BMNH 70.2.17.22	Ep				B_0	B_0	B_0	B_0	B_0	B_0	B_0	B_0	B_0	B_0	B_0	B_4	B_4	B_4	B_4	B_4	B_4	B_4	B_4	B_4	B_4	B_4	B_4	B_4	B_4	B_3	DB_3	DB_5	DB_5
SL ca 70 mm	R			S		B	B	B	B	B	B	B	B	B	B	B	B	B	B	B	B	B	B	B	B	B	B	B	B	B	B	B	B
Clupeichthys	En	L	-	L_1	L_1	B_1	B_1	B_1	B_1	B_1	B_1	B_1	B_1	B_1	B_1	B_1	B_1	B_1	B_1	B_1	vB_1	vB_1	vB_1	vB_1	vB_1	vB_1	vB_1	vB_1	vB_1	vB_1	vB_1	vB_1	vB_1
bleekeri	Ec	B		cB_1	cB_2	cF_4	cF_4	cF_4	cF_4	cF_4	cF_4	cF_4	cF_4	cF_4	cF_4	cF_4	cF_4	cF_4	cF_4	cF_4	cF_4	cB_2	cB_2	cB_2	cB_2	L_2	L_2	L_2					
BMNH 79.3.21.105	Ep					D_0	D_0	D_0	D_0	D_0	D_0	D_0	D_0	D_0	D_0	D_0	D_0	D_0	D_0	D_0	DB_3	DB_3	DB_3	DB_3	DB_3	DB_3	DB_5	DB_5	DB_5	DB_5	DB_5	DB_5	DB_5
SL 58 mm	R			S		B	B	B	B	B	B	B	B	B	B	B	B	B	B	B	B	B	B	B									
Jenkinsia	En		-			F_1	F_1	F_1	F_1	F_1	F_1	F_1	F_1	F_1	F_1	F_1	F_1	F_1	B_1	B_1	B_1	B_1	B_1	vB_1	vB_1	vB_1	vB_1	vB_1	vB_1	vB_1	vB_1	vB_1	vB_1
stolifera	Ec	cB		cB_2	cB_2	cF_4	cF_4	cF_4	cF_4	cF_4	cF_4	cF_4	cB_3	cB_3	cB_3	cB_3	cB_3	cB_3	cB_3	cB_3	cB_3	cB_3	cB_2	cB_2	cB_2	B_2							
BMNH 65.7.12.46	Ep				D_0	D_0	D_0	D_0	D_0	D_0	D_0	D_0	D_0	D_0	D_0	D_0	D_0	D_0	D_0	D_0	D_0	D_3	D_3	B_5D_0	DB_5	DB_5	DB_5	DB_5	DB_5	DB_5	DB_5	DB_5	DB_5
SL 34 mm	R			S		B	B	B	B	B	B	B	B	B	B	B	B	B	B	B	B	B	B	B	B	B					
Anchoa	Ep	B	B	B_1	B_1	B_2	B_2	B_2	B_2	B_2	B_2	vB_2	vB_2	vB_2	vB_2	vB_2	vB_2	vB_2	vB_2	vB_1	vB_1	vB_1	vB_1	vB_1	vB_1	vB_1	vB_1	vB_1	vB_1	vB_1	vB_1	vB_1	vB_1
delicatissima	Ep	B	B	cB_2	cB_2	cB_3	cB_3	cB_3	cB_3	cB_3	cB_3	cB_3	cB_3	cB_3	cB_3	cB_3	cB_3	B_3	B_3	B_3	B_3	B_3	B_2	B_2	B_2	L_2	L_2	L_2	L_2	L_2	L_2	L_2	L_2
W50V26	Ep				D_0	DB_0	DB_4	DB_4	DB_4	DB_4	DB_4	DB_4	DB_4	DB_4	DB_4	DB_4	DB_4	DB_4	DB_4	DB_3	DB_3	DB_3	DB_5	DB_5	DB_5	DB_5	DB_5	DB_5	DB_5	DB_5	DB_5	DB_5	DB_5
SL 64 mm	Ep			S		B	B	B	B	B	B	B	B	B	B	B	B	B	B	B	B												
Chirocentrus	En	BB	+	F_1	F_1	B_1	vB_1	vB_1	vB_1	vB_1	vB_1	vB_1	vB_1	vB_1	vB_1	vB_1	vB_1	vB_1	vB_1	vB_1	vB_1	vB_1	vB_1	vB_1	vB_1	vB_1	vB_1	vB_1	vB_1	vB_1	vB_1	vB_1	vB_1
dorab	Ec	BB		B_1	F_2	F_4	F_4	F_4	F_4	F_4	F_4	F_4	B_3	B_3	B_3	B_3	B_2	B_2	B_2	B_2	B_2	B_2	B_2	B_2	B_2	B_2	B_2	B_2	B_2	B_2	B_2	B_2	B_2
BMNH uncat.	Ep	<DD		D_0	D_0	D_0	D_0	D_0	D_0	D_0	D_0	D_0	D_0	D_0	D_0	D_0	B_3D_0	B_3D_0	B_3D_0	B_3D_0	DB_3	DB_3	DB_3	DB_3	DB_3	DB_3	DB_3	DB_3	DB_3	DB_3	DB_3	DB_3	DB_3
SL 147 mm	R				S	SB	B	B	B	B	B	B	B	B	B	B	B	B	B	B	B	B	B	B	B	B	B	B	B	B	B	B	B

31 32 33 34 35 36 37 38 39 40 41 42 43 44 45 46 47 48 49 50 51 52 53 54 55 56 57 58 59 60 61 62 63 64 65 66 67 68 69 70 71 72

$L_1\ L_1\ L_1\ L_1\ L_1\ L_1\ L_1\ L_1$
$L_3\ L_3\ L_5\ L_5\ L_5\ L_2\ L_2\ L_2\ L_2\ L_2\ L_2\ L_2\ L_2\ L_2\ L_2\ L_2\ L_2$ **PU10**
$L_4\ L_5\ L_5\ L_5\ L_5\ L_5\ L_5\ L_5$
$\underline{B}$

$B_1\ B_1\ B_1\ B_1\ B_1$
$L_2\ L_2\ L_2\ L_2\ L_2\ L_2\ L_2\ L_2\ L_2\ L_2\ L_2\ L_2\ L_2\ L_2\ L_2\ L_2\ L_2\ L_2\ L_2\ L_2$ **PU34**
$L_5\ L_5\ L_5\ L_5\ L_5\ L_5\ L_5\ L_5\ L_5\ L_5\ L_5\ L_5\ L_5\ L_5\ L_5\ L_5\ L_5\ L_5\ L_5\ L_5\ L_5\ L_5\ L_5\ L_5\ L_5\ L_5$

$B_1\ B_1\ B_1\ B_1$
$L_2\ L_2\ L_2\ L_2\ L_2\ L_2\ L_2\ L_2\ L_2\ L_2\ L_2\ L_2\ L_2$ **PU21**
$B_5\ B_5\ B_5$

M M M M M M M M M M M M M M M M M M M M M M M M M M M M M M M M M M M
L_2 **PU4**
M M M M M M M M M M M M M M M M M M M M M M M M M M M M M M M M

${}_{V}B_1\ {}_{V}B_1\ {}_{V}B_1\ {}_{V}B_1\ {}_{V}B_1\ {}_{V}B_1\ {}_{V}B_1\ {}_{V}B_1\ {}_{V}B_1\ {}_{V}B_1\ {}_{V}B_1\ {}_{V}B_1\ {}_{V}B_1\ V_0\ V_0\ V_0\ V_0\ V_0\ V_0\ V_0\ V_0\ V_0\ V_0\ V_0\ V_0\ V_0\ V_0$
$L_2\ L_2\ L_2\ L_2\ L_2\ L_2\ L_2\ L_2\ L_2\ L_2\ L_2\ L_2\ L_2\ L_2\ L_2\ L_2\ L_2\ L_2\ L_2\ L_2$ **PU3**
${}_{D}B_0\ {}_{D}B_4\ {}_{D}B_4\ {}_{D}B_4\ {}_{D}B_4\ {}_{D}B_4\ {}_{D}B_4\ {}_{D}B_4\ {}_{D}B_4\ {}_{D}B_4\ {}_{D}B_4\ {}_{D}B_4\ D_0\ D_0\ D_0\ D_0\ D_0\ D_0\ D_0\ D_0\ D_0\ D_0\ D_0\ D_0\ D_0\ D_0\ D_0$
$\underline{B}$ - $\underline{B}\ \underline{B}\ \underline{B}\ \underline{B}\ \underline{B}\ \underline{B}\ \underline{B}\ \underline{B}\ \underline{B}\ \underline{B}\ \underline{B}$

$F_1\ {}_{V}B_1\ {}_{V}B_1\ {}_{V}B_1\ {}_{V}B_1\ {}_{V}B_1\ V_1\ V_1\ V_1\ V_1\ V_1\ V_1\ V_1\ V_1\ V_1\ V_0\ V_0\ V_0\ V_0\ V_0\ V_0\ V_0\ V_0\ V_0\ V_0\ V_0\ V_0\ V_0\ V_0\ V_0\ V_0\ V_0\ V_0$
$B_2\ B_2\ B_2\ B_2\ B_2\ B_2\ B_2\ B_2\ B_2\ B_2\ B_2\ B_2\ {}_{D}B_2\ B_2\ B_2\ B_2\ B_2\ B_2\ L_2\ L_2\ L_2\ L_2\ L_2\ L_2\ L_2\ L_2$ **PU4**
${}_{D}B_4\ {}_{D}B_4\ {}_{D}B_4\ {}_{D}B_4\ {}_{D}B_4\ {}_{D}B_4\ {}_{D}B_4\ {}_{D}B_4\ {}_{D}B_4\ {}_{D}B_4\ {}_{D}B_4\ D_0\ D_0\ D_0\ D_0\ D_0\ D_0\ D_0\ D_0\ D_0\ D_0\ D_0\ D_0\ D_0\ D_0\ D_0\ D_0\ D_0\ D_0\ D_0$
$\underline{B}\ \underline{B}\ \underline{B}\ \underline{B}\ \underline{B}\ \underline{B}\ \underline{B}\ \underline{B}\ \underline{B}\ \underline{B}$

$B_1\ B_1\ B_1\ {}_{V}B_1\ {}_{V}B_1\ {}_{V}B_1\ {}_{V}B_1\ {}_{V}B_1\ {}_{V}B_1\ {}_{V}B_1\ {}_{V}B_1\ {}_{V}B_1\ {}_{V}B_1\ {}_{V}B_1\ {}_{V}B_1\ {}_{V}B_1\ {}_{V}B_1\ {}_{V}B_1\ {}_{V}B_1\ {}_{V}B_0\ {}_{V}B_0\ {}_{V}B_0\ {}_{V}B_0\ {}_{V}B_0\ {}_{V}B_0\ {}_{V}B_0\ V_0\ V_0\ V_0\ V_0\ V_0\ V_0\ V_0$
$L_4\ L_4\ L_4\ L_4\ L_4\ L_4\ L_3\ L_3\ L_3\ L_3\ L_3\ L_3\ L_3\ L_2\ L_2\ L_2\ L_2\ L_2\ L_2\ L_2\ L_2\ L_2\ L_2$ **PU8**
$B_4\ B_4\ B_4\ B_4\ B_4\ B_4\ B_4\ B_4\ B_4\ B_4\ {}_{D}B_4\ {}_{D}B_4\ {}_{D}B_4\ {}_{D}B_4\ {}_{D}B_4\ {}_{D}B_4\ {}_{D}B_4\ {}_{D}B_4\ {}_{D}B_4\ {}_{D}B_5\ {}_{D}B_5\ {}_{D}B_0\ {}_{D}B_0\ {}_{D}B_0\ D_0\ D_0\ D_0\ D_0\ D_0\ D_0\ D_0\ D_0\ D_0$
$\underline{B}\ \underline{B}$

$B_1\ B_1\ B_1\ B_1\ B_1\ {}_{V}B_1\ V_0\ V_0\ V_0\ V_0\ V_0\ V_0\ V_0\ V_0\ V_0\ V_0$
$L_3\ L_3\ L_3\ L_3\ L_2\ L_2\ L_2\ L_2\ L_2\ L_2\ L_2\ L_2\ L_2\ L_2\ L_2\ L_2\ L_2\ L_2\ L_2\ L_2\ L_2\ L_2\ L_2\ L_2\ L_2\ L_2\ L_2\ L_2\ L_2\ L_2\ L_2\ L_2\ L_2\ L_2\ L_2\ L_2\ L_2$ **PU6**
$B_4\ B_4\ B_4\ B_4\ B_4\ B_4\ B_4\ B_3\ B_3\ B_3\ B_3\ B_3\ {}_{D}B_3\ {}_{D}B_3\ {}_{D}B_3\ {}_{D}B_3\ {}_{D}B_5\ {}_{D}B_5\ {}_{D}B_5\ {}_{D}B_5\ {}_{D}B_5\ {}_{D}B_5\ {}_{D}B_5\ {}_{D}B_5\ {}_{D}B_5\ {}_{D}B_5\ D_0\ D_0\ D_0\ D_0\ D_0\ D_0\ D_0\ D_0\ D_0\ D_0\ D_0$
$\underline{B}\ \underline{B}$

$B_1\ B_1\ B_1\ B_1\ B_1\ B_1\ B_1\ B_1\ B_1\ B_1\ B_1\ B_1\ {}_{V}B_1\ {}_{V}B_1\ {}_{V}B_1\ {}_{V}B_1\ {}_{V}B_1\ {}_{V}B_1\ V_0\ V_0\ V_0\ V_0\ V_0\ V_0\ V_0\ V_0\ V_0\ V_0\ V_0\ V_0\ V_0\ V_0\ V_0\ V_0$
$L_2\ L_2\ L_2\ L_2\ L_2\ L_2\ L_2\ L_2\ L_2\ L_2\ L_2\ L_2\ L_2\ L_2\ L_2\ L_2\ L_2\ L_2\ L_2\ L_2\ L_2\ L_2$ **PU9**
$L_4\ L_4\ L_4\ L_3\ L_3\ L_3\ L_3\ L_3\ L_3\ L_3\ L_3\ L_3\ B_3\ B_3\ B_3\ {}_{D}B_3\ B_5\ {}_{D}B_5\ {}_{D}B_5\ {}_{D}B_5\ {}_{D}B_5\ {}_{D}B_5\ D_0\ D_0\ D_0\ D_0\ D_0\ D_0\ D_0\ D_0\ D_0\ D_0\ D_0$
$\underline{B}\ \underline{B}\ \underline{B}\ \underline{B}\ \underline{B}\ \underline{B}\ \underline{B}\ \underline{B}\ \underline{B}\ \underline{B}\ \underline{B}\ \underline{B}\ \underline{B}\ \underline{B}$ L$\underline{B}$

${}_{V}B_1\ V_1\ {}_{V}B_1\ V_1\ V_1\ V_1\ V_1\ V_1\ V_1\ V_1\ V_1\ V_1\ V_1$ V
$L_3\ L_3\ L_3\ L_3\ L_3\ L_3\ L_3\ L_3\ L_3\ L_3\ L_3\ L_3\ L_3\ L_3\ L_3\ L_3\ L_3\ L_3\ L_3\ L_3\ L_3\ L_3\ L_3\ L_3\ L_3\ L_3\ L_3\ L_3\ L_3\ L_3\ L_3\ L_3\ L_3\ L_3\ L_3\ L_3\ L_3\ L_3\ L_3\ L_3\ L_2$ L
$L_4\ L_4\ L_4\ L_4\ L_4\ L_4\ L_4\ L_4\ L_4\ L_4\ L_4\ L_4\ L_4\ L_4\ L_4\ L_4\ L_4\ L_3\ L_3\ L_3\ L_3\ D_3\ D_3\ D_3\ D_3\ D_3\ D_3\ D_3\ D_3\ D_3\ D_3\ D_3\ D_3\ D_5\ D_5$ D
B B B B B B B B B B B B B B B B B B B B B B B L B L L L L L L L L L L

${}_{V}B_1\ V_1$ V
$L_3\ L_3\ L_3\ L_3\ L_3\ L_3\ L_3\ L_3\ L_3\ L_3\ L_3\ L_3\ L_3\ L_3\ L_3\ L_3\ L_3\ L_3\ L_3\ L_2\ L_2\ L_2\ L_2\ L_2\ L_2\ L_2\ L_2\ L_2\ L_2\ L_2\ L_2\ L_2\ L_2\ L_2\ L_2\ L_2\ L_2\ L_2\ L_2\ L_2\ L_2$ L
$L_3\ L_3\ L_3\ L_3\ L_3\ L_3\ L_3\ L_3\ L_3\ L_3\ B_3\ B_3\ {}_{D}B_3\ D_3\ {}_{D}B_3\ {}_{D}B_3\ {}_{D}B_3\ {}_{D}B_3\ {}_{D}B_5\ {}_{D}B_5\ {}_{D}B_5\ {}_{D}B_5\ {}_{D}B_5\ {}_{D}B_5\ {}_{D}B_5\ {}_{D}B_5\ {}_{D}B_5\ {}_{D}B_5\ {}_{D}B_5\ {}_{D}B_5\ {}_{D}B_5\ {}_{D}B_5\ {}_{D}B_5\ {}_{D}B_5\ D_5\ D_5\ {}_{D}B_5\ D_5\ D_5\ {}_{D}B_5$ D
L L L L L L L L L L L L L

$B_1\ B_1\ B_1\ B_1\ B_1\ B_1\ B_1\ B_1\ B_1\ B_1\ B_1\ B_1\ B_1\ B_1\ B_1\ B_1\ B_1\ B_1\ B_1\ B_1\ B_1\ B_1\ B_1\ B_1\ B_1\ B_1\ B_1\ B_1\ B_1\ B_1\ B_1\ B_1\ B_1\ B_1\ B_1\ B_1\ B_1\ B_1\ B_1\ B_1\ B_1$ B
$L_3\ L_3\ L_3\ L_3\ L_3\ L_3\ L_3\ L_3\ L_3\ L_3\ L_3\ L_3\ L_3\ L_2\ L_2\ L_2\ L_2\ L_2\ L_2\ L_2\ L_2\ L_2\ L_2\ L_2\ L_2\ L_2\ L_2\ L_2\ L_2\ L_2\ L_2\ L_2\ L_2\ L_2\ L_2\ L_2\ L_2\ L_2\ L_2\ L_2$ L
$L_3\ L_3\ L_3\ L_3\ L_3\ B_3\ B_3\ B_3\ B_3\ B_3\ B_3\ B_3\ B_3\ B_5\ B_5\ B_5\ B_5\ B_5\ B_5\ B_5\ B_5\ B_5\ B_5\ B_5\ B_5\ B_5\ B_5\ B_5\ B_5\ B_5\ B_5\ B_5\ B_5\ B_5\ B_5\ B_5\ B_5\ B_5\ B_5\ B_5$ B
B B B B B B B

$V_0\ V_0\ V_0\ V_0\ V_0\ V_0\ V_0\ V_0$ **PU1**
$D_0\ D_0\ D_0\ D_0\ D_0\ D_0\ D_0\ D_0$

${}_{V}B_1\ {}_{V}B_1\ {}_{V}B_1\ {}_{V}B_1\ {}_{V}B_1\ {}_{V}B_1\ {}_{V}B_1\ {}_{V}B_1\ {}_{V}B_1\ {}_{V}B_1\ {}_{V}B_1\ {}_{V}B_1\ {}_{V}B_1\ {}_{V}B_1\ {}_{V}B_1\ V_0\ V_0\ V_0\ V_0\ V_0\ V_0\ V_0\ V_0\ V_0\ V_0$
$B_2\ B_2\ B_2\ B_2\ L_2\ L_2\ L_2\ L_2\ L_2\ L_2\ L_2\ L_2\ L_2\ L_2\ L_2\ L_2\ L_2$ **PU3**
${}_{D}B_5\ {}_{D}B_5\ {}_{D}B_5\ {}_{D}B_5\ {}_{D}B_5\ {}_{D}B_5\ {}_{D}B_5\ {}_{D}B_5\ {}_{D}B_5\ {}_{D}B_5\ {}_{D}B_5\ {}_{D}B_5\ {}_{D}B_5\ {}_{D}B_5\ {}_{D}B_5\ D_0\ D_0\ D_0\ D_0\ D_0\ D_0\ D_0\ D_0$
$\underline{B}\ \underline{B}\ \underline{B}\ \underline{B}\ \underline{B}\ \underline{B}\ \underline{B}$

${}_{V}B_1\ {}_{V}B_1\ {}_{V}B_1\ V_0\ V_0\ V_0\ V_0\ V_0$ **PU1**
${}_{D}B_5\ D_0\ D_0\ D_0\ D_0\ D_0\ D_0\ D_0$

${}_{V}B_1\ {}_{V}B_1\ V_0\ V_0\ V_0\ V_0\ V_0$ **PU2**
${}_{D}B_5\ {}_{D}B_5\ {}_{D}B_5\ D_0\ D_0\ D_0$

${}_{V}B_1\ {}_{V}B_1\ V_0\ V_0\ V_0\ V_0\ V_0\ V_0\ V_0\ V_0$
$L_2\ L_2\ L_2\ L_2$ **PU1**
${}_{D}B_5\ {}_{D}B_5\ {}_{D}B_5\ D_0\ D_0\ D_0\ D_0\ D_0\ D_0\ D_0$

${}_{V}B_1\ {}_{V}B_1\ {}_{V}B_1\ {}_{V}B_1\ {}_{V}B_1\ {}_{V}B_1\ {}_{V}B_1\ {}_{V}B_1\ {}_{V}B_1\ {}_{V}B_1\ {}_{V}B_1\ {}_{V}B_1\ {}_{V}B_1\ {}_{V}B_1\ {}_{V}B_1\ {}_{V}B_1\ {}_{V}B_1\ {}_{V}B_1\ {}_{V}B_1\ V_1\ V_1\ V_1\ V_1\ V_1\ V_1\ V_1\ V_1\ V_1\ V_0\ V_0\ V_0\ V_0\ V_0\ V_0\ V_0\ V_0\ V_0\ V_0\ V_0\ V_0$
$B_2\ B_2\ B_2\ B_2\ B_2\ B_2\ B_2\ B_2\ B_2\ B_2\ B_2\ B_2\ B_2\ B_2\ B_2\ B_2\ B_2\ B_2\ B_2\ B_2\ B_2\ B_2\ B_2\ B_2\ B_2\ B_2\ B_2\ B_2\ B_2\ B_2\ L_2\ L_2\ L_2\ L_2$ **PU1**
${}_{D}B_3\ {}_{D}B_5\ D_5\ {}_{D}B_5\ D_5\ D_5\ {}_{D}B_5\ D_5\ D_5\ D_5\ D_0\ D_0\ D_0\ D_0\ D_0\ D_0\ D_0\ D_0\ D_0\ D_0\ D_0$
$\underline{B}\ \underline{B}\ \underline{B}\ \underline{B}\ \underline{B}\ \underline{B}\ \underline{B}\ \underline{B}\ \underline{B}\ \underline{B}\ \underline{B}\ \underline{B}$

73 // 80 81 82 83 84 85 86 // 110 111 112 // 129 130 // 134 135 // 140 141 142 143 144 145 146 147 148 149 150 151 152

V_1 // V_1 V_1 V_1 V_1 V_1 V_1 V_1 // V_1 V_1 V_1 // V_1 V_0 // V_0 V_0 // V_0 V_0 V_0 V_0 V_0 V_0
L_2 // L_2 L_2 L_2 L_2 L_2 L_2 L_2 // L_2 L_2 L_2 // L_2 L_2 // L_2 L_2 // L_2 **PU48**
D_5 // D_5 D_5 D_5 D_5 D_5 D_5 D_5 // D_5 D_5 D_5 // D_5 D_5 // D_5 D_0 // D_0 D_0 D_0

V_1 // V_1 V_1 V_1 V_1 V_1 V_1 V_1 // V_1 V_1 V_1 // V_0 V_0 // V_0 V_0 // V_0 V_0 V_0 V_0 V_0 V_0 V_0 V_0 V_0 V_0
L_2 // L_2 L_2 L_2 L_2 L_2 L_2 L_2 // L_2 L_2 // // // **PU75**
D_5 // D_5 D_5 D_5 D_5 D_5 D_5 D_5 // D_5 D_5 D_5 // D_5 D_5 // D_5 D_5 // D_5 D_5 D_5 D_5 D_5 D_5 D_5 D_5 D_5 D_5 D_5 D_5

B_1 // B_1 B_1 B_1 B_1 B_1 B_1
L_2 // L_2 **PU25**
B_5 // B_5 B_5 B_5 B_5 B_5 B_5